Reveillac,
Jean-Michel,
author.
Musical sound
effects

SEP 1 0 2018

Musical Sound Effects

Musical Sound Effects

Analog and Digital Sound Processing

Jean-Michel Réveillac

WILEY

First published 2018 in Great Britain and the United States by ISTE Ltd and John Wiley & Sons, Inc.

Apart from any fair dealing for the purposes of research or private study, or criticism or review, as permitted under the Copyright, Designs and Patents Act 1988, this publication may only be reproduced, stored or transmitted, in any form or by any means, with the prior permission in writing of the publishers, or in the case of reprographic reproduction in accordance with the terms and licenses issued by the CLA. Enquiries concerning reproduction outside these terms should be sent to the publishers at the undermentioned address:

ISTE Ltd
27-37 St George's Road
London SW19 4EU
UK

www.iste.co.uk

John Wiley & Sons, Inc.
111 River Street
Hoboken, NJ 07030
USA

www.wiley.com

© ISTE Ltd 2018

The rights of Jean-Michel Réveillac to be identified as the author of this work have been asserted by him in accordance with the Copyright, Designs and Patents Act 1988.

Library of Congress Control Number: 2017954678

British Library Cataloguing-in-Publication Data
A CIP record for this book is available from the British Library
ISBN 978-1-78630-131-4

Contents

Foreword xi

About this Book xiii

Introduction xvii

Chapter 1. Notes on the Theory of Sound 1

    1.1. Basic concepts 1
        1.1.1. What is sound? 1
        1.1.2. Intensity 4
        1.1.3. Sound pitch 7
        1.1.4. Approaching the concept of timbre 8
    1.2. The ears. 9
        1.2.1. How our ears work 9
        1.2.2. Fletcher–Munson curves. 14
        1.2.3. Auditory spatial awareness 15
    1.3. The typology of sounds. 21
        1.3.1. Sounds and periods. 21
        1.3.2. Simple sounds and complex sounds 22
    1.4. Spectral analysis 24
        1.4.1. The sound spectrum 24
        1.4.2. Sonogram and spectrogram 26
    1.5. Timbre. 27
        1.5.1. Transient phenomena 27
        1.5.2. Range. 28
        1.5.3. Mass of musical objects 30
        1.5.4. Classification of sounds 31
    1.6. Sound propagation. 32
        1.6.1. Dispersion 32

1.6.2. Interference	33
1.6.3. Diffraction	35
1.6.4. Reflection	37
1.6.5. Reverberation (reverb)	39
1.6.6. Absorption	39
1.6.7. Refraction	39
1.6.8. The Doppler effect	40
1.6.9. Beats	40
1.7. Conclusion	41
Chapter 2. Audio Playback	**43**
2.1. History	44
2.2. Dolby playback standards and specifications	48
2.2.1. Dolby Surround encoding and decoding	48
2.2.2. Dolby Stereo	49
2.2.3. Dolby Surround	50
2.2.4. Dolby Surround Pro-Logic	50
2.2.5. Dolby DIGITAL AC-3	51
2.2.6. Dolby Surround EX	52
2.2.7. Dolby Surround Pro-Logic II	52
2.2.8. Dolby Digital Plus	54
2.2.9. Dolby TrueHD	54
2.2.10. Dolby Atmos	55
2.3. DTS encodings	55
2.3.1. DTS	56
2.3.2. DTS Neo 6	56
2.3.3. DTS ES 6.1	57
2.3.4. DTS 96/24	57
2.3.5. DTS HD Master Audio	57
2.3.6. DTS X	58
2.4. Special encodings	58
2.5. SDDS	59
2.6. THX certification	59
2.6.1. THX select and ultracertification	61
2.6.2. THX Ultra 2 certification	61
2.7. Multichannel audio recording	62
2.8. Postproduction and encoding	63
2.9. Multichannel music media: DVD-Audio and SACD	65
2.9.1. DVD-Audio	65
2.9.2. Super Audio CD	67
2.9.3. Comparison of CDs, SACDs and DVD-Audios	69
2.10. Conclusion	69

Chapter 3. Types of Effect ... 71

3.1. Physical appearance ... 71
   3.1.1. Racks ... 72
   3.1.2. Pedals ... 74
   3.1.3. Software plugins ... 77
3.2. Audio processing ... 78
3.3. Conclusion ... 80

Chapter 4. Filtering Effects ... 81

4.1. Families of filtering effects ... 81
4.2. Equalization ... 84
   4.2.1. Frequency bands and ranges ... 84
   4.2.2. Types of equalizer ... 86
   4.2.3. Examples of equalizers ... 91
   4.2.4. Tips for equalizing a mix ... 94
4.3. Wah-wah ... 97
   4.3.1. History ... 97
   4.3.2. Theory ... 99
   4.3.3. Auto-wah ... 100
   4.3.4. Examples of wah-wah pedals ... 101
4.4. Crossover ... 102
4.5. Conclusion ... 104

Chapter 5. Modulation Effects ... 105

5.1. Flanger ... 105
   5.1.1. History ... 105
   5.1.2. Theory and parameters ... 107
   5.1.3. Models of flanger ... 110
5.2. Phaser ... 111
   5.2.1. Examples of phasers ... 113
5.3. Chorus ... 115
   5.3.1. Examples of chorus ... 116
5.4. Rotary, univibe or rotovibe ... 117
   5.4.1. History ... 118
   5.4.2. Theoretical principles ... 120
   5.4.3. Leslie speakers ... 122
   5.4.4. Examples of rotary or univibe pedals ... 123
   5.4.5. Leslie speakers and sound recording ... 125
5.5. Ring modulation ... 127
   5.5.1. Theoretical principles ... 127
   5.5.2. Examples of ring modulators ... 129
5.6. Final remarks ... 130

Chapter 6. Frequency Effects ... 131

6.1. Vibrato ... 131
    6.1.1. Theoretical principles ... 132
    6.1.2. Settings ... 132
    6.1.3. Examples of vibrato ... 133
6.2. Transposers ... 135
    6.2.1. Octaver ... 135
    6.2.2. Pitch shifter ... 137
    6.2.3. Harmonizer ... 139
    6.2.4. Auto-Tune ... 142
6.3. Conclusion ... 154

Chapter 7. Dynamic Effects ... 155

7.1. Compression ... 156
    7.1.1. History ... 156
    7.1.2. Parameters of compression ... 156
    7.1.3. Examples of compressors ... 163
    7.1.4. Multiband compressors ... 166
    7.1.5. Guidelines for configuring a compressor ... 169
    7.1.6. Parallel compression ... 170
    7.1.7. Serial compression ... 171
    7.1.8. Compression with a sidechain ... 171
    7.1.9. Some basic compression settings ... 172
    7.1.10. Synchronizing the compressor ... 175
    7.1.11. Using a compressor as a limiter ... 175
7.2. Expanders ... 178
    7.2.1. Parameters ... 178
    7.2.2. Examples of software expanders ... 179
7.3. Noise gates ... 180
    7.3.1. Parameters ... 180
    7.3.2. Examples of noise gates ... 182
    7.3.3. Configuring noise gates ... 184
7.4. De-essers ... 184
    7.4.1. Principle of a de-esser ... 184
    7.4.2. Examples of de-essers ... 185
    7.4.3. Replacing a de-esser with an equalizer and a compressor ... 186
    7.4.4. Configuring a de-esser ... 186
7.5. Saturation ... 187
    7.5.1. Fuzz ... 187
    7.5.2. Overdrive ... 188
    7.5.3. Distortion ... 188
    7.5.4. Examples of equipment dedicated to creating saturation ... 189

7.6. Exciters and enhancers 192
   7.6.1. Examples of exciters 193
   7.6.2. Using a sound enhancer 195
7.7. Conclusion 195

Chapter 8. Time Effects 197

8.1. Reverb 197
   8.1.1. Theoretical principles 197
   8.1.2. History 200
   8.1.3. Principles 208
   8.1.4. Reverb configuration 219
   8.1.5. Recording the IR and deconvolution 227
   8.1.6. Studio mixing and reverb 240
8.2. Delay 243
   8.2.1. History 243
   8.2.2. Types of delay 244
   8.2.3. Tips for using delays in the studio 251
8.3. Conclusion 255

Chapter 9. Unclassifiables 257

9.1. Combined effects 257
   9.1.1. Fuzzwha 257
   9.1.2. Octafuzz 258
   9.1.3. Shimmer 259
9.2. Tremolo 262
   9.2.1. History 262
   9.2.2. Examples of tremolos 264
9.3. Sound restoration tools 266
   9.3.1. Declickers 266
   9.3.2. Decracklers 267
   9.3.3. Denoisers 267
   9.3.4. Declippers 267
   9.3.5. Debuzzers 268
   9.3.6. Examples of restoration tools 268
   9.3.7. Final remarks on sound restoration 271
9.4. Loopers 271
   9.4.1. Looper connections 272
   9.4.2. Examples of looper pedals 274
9.5. Time stretching 275
9.6. Resampling 276
9.7. Spatialization effects 277
9.8. Conclusion 278

Conclusion. 279

Appendices 283

Appendix 1. 285

Appendix 2. 295

Appendix 3. 299

Appendix 4. 313

Glossary 319

Bibliography 327

Index 337

Foreword

What will music look like in the future?

If you are wondering about the building blocks of the music of tomorrow, or if you wish to understand them, this book will prove a valuable "toolbox". Jean Michel Réveillac introduces us to the myriad of sounds effects of the world of music. He weaves implicit threads between the art of sound and the history of science, allowing us to appreciate how digital effects influence the typology and morphology of physical waves.

The materials that will be used by the sound architects of our future are playback images, colors, matter and expressions. The possibilities of sound effect processing afford a glimpse into a universe of infinite and unsuspected dimensions of human hearing. By chronologically and thematically exploring these physical phenomena, Jean Michel Réveillac not only reveals the path, but delicately retraces the resources and techniques of the scientific, philosophical and sociological tradition in which the history of these technologies is steeped. The transmitted codes of electronic music have made space for sound objects, creating a certain anatomy of sound, and many other musical movements have also drawn inspiration from various applied effects.

The breakthrough transition to a fully digital world is uprooting the traditions of analog processing. The increasing capabilities of DAWs[1] and 5.1 multichannel mixing are challenging the old trades and customs of applied effects. Music itself is a canvas for composers to shape according to their desires of expression by sound narration, and digital audio tools are a palette for virtual modeling. Today, sound occupies a permanent place within our environment. Jean Michel Réveillac revisits the original historical approaches enriched with the knowledge of digital audio

---

1 Digital Audio Workstations.

processing. By broadening its field of investigation, the history of sound effects has carved its role in how we understand the challenges of modern sound. This book upholds the quality of its scientific roots and the relevance of its forays into audio research. After exploring the wealth of knowledge on musical processing present in this book, we are left with no doubt that a summary of modern research into the history of sound effects was sorely needed.

As if set in stone, the chosen approach conveys the timelessness of mixing techniques and colors, forging an eternal record of the methodology of recent years. Leaving us with the feeling that this was just the beginning of a journey to the heart of a perpetually expanding culture. Many of the sound effects presented in this book are currently extremely popular, with multiple areas of application. Chapter by chapter, the singular vision of a constantly evolving landscape of audio effects gradually emerges. With great passion, the author retraces the greatest events in the history of sound effects and the key theoretical ideas that accompanied them, and offers his own thoughts on the origin of this universe that has drawn him ever deeper, as well as the impact and future of technological advancements that will allow everyone to leave their own mark on the evolution of sound.

Léo PAOLETTI
"Leo Virgile"
Composer and audio designer

About this Book

If you are wondering whether this book is for you, how it is put together and organized, what it contains and which conventions we will use, you are in the right place. Everything will be explained here.

Target audience

This book is intended for anybody who is passionate about sound – amateurs or professionals who love sound recording, mixing or playback, and, of course, musicians, performers and composers.

The topics discussed in a few sections require some basic knowledge of the principles of general computing and digital audio.

You need to be familiar with operating systems (paths, folders and directories, files, filenames, file extensions, copying, dragging, etc.) and you need to know how to use a digital audio editor like Adobe Audition, Steinberg Wavelab, Magix Sound Forge, Audacity, etc., or a DAW (Digital Audio Workstation) such as Avid Pro Tools, Magix Samplitude, Sonar Cakewalk, Apple Logic Pro and so on.

Organization and contents

This book has nine chapters:

– Notes on the Theory of Sound;

– Audio Playback;

– Types of Effects;

– Filtering Effects;

– Modulation Effects;

– Frequency Effects;

– Dynamic Effects;

– Time Effects;

– Unclassifiables.

Each of these chapters can be read separately. Some concepts depend on concepts from other chapters, but references are given wherever they are needed. The first chapter, dedicated to the theory of sound, is slightly different. It provides the basic foundations needed to understand each of the other chapters.

If you are new to the scene, I highly recommend reading the first chapter. The rest of the book will be easier to understand.

Even if you are not, it might still surprise you with a few new ideas.

The conclusion, unsurprisingly, attempts to give an overview of the current state of the world of sound effects, and how they might continue to develop in future.

Appendices 1–4 discuss a few extra ideas and reminders in the following order:

– Distortion;

– Classes of Amplifiers;

– Basics with Max/MSP;

– Multieffect Racks.

A bibliography and a list of Internet links can be found at the end of the book.

There is also a glossary explaining some of the logos, acronyms and terminology specific to sound effects, sound recording, mixing and playback.

Conventions

The following formatting conventions are used throughout the book:

– *italics*: indicates the first time that an important term is used. For example, this could be one of the words explained in the glossary at the end of the book, mathematical terms, comments, equations, formulas or variables;

– *(italics)*: terms written in languages other than English;

– CAPS: names of windows, icons, buttons, folders or directories, menus or submenus. This also includes elements, options or controls in the windows of a software program.

Additional remarks are identified by the keyword.

REMARK.– These comments supplement the explanations given in the main body of the text.

All figures and tables have explanatory captions.

Vocabulary and definitions

Like all specialized techniques, audio sound effects have their own vocabulary. Some of the words, acronyms, abbreviations, logos and proper nouns may not be familiar to all readers. The glossary mentioned earlier should help with these terms.

Acknowledgments

I would especially like to thank the team at ISTE and my editor, Chantal Ménascé, for placing their trust in me. I am greatly indebted to the composer and musician Léo Paoletti, known by his stage name "Leo Virgile," for the time, attention and interest he has granted me, as well as for writing the foreword to this book.

Finally, I would like to thank my wife, Vanna, for never ceasing to support me as I wrote these pages.

Introduction

Sound effects have always fascinated me. I remember my amazement and disbelief as a child when I heard the sound of my voice echoing for the first time when hiking on a mountain in the Alps. How is this possible?

My father, who was with me at the time, tried to give me a simple explanation of the phenomenon, but I could not really understand what he meant, and for years it remained a mystery to me.

Maybe this was what triggered my passion for listening, observing and understanding how sounds drift and wander within their natural environments, later leading to my interest in how these sounds can be artificially processed.

Today, sound effects are most commonly used for shows, movies and musical production. With the proliferation of microcomputers and electronics, creating sound effects has never been easier. New tools and software can do things that we would have struggled to imagine just a few years ago.

The idea of this book arises from the many questions that people have asked me when they visit my studio, over the course of my university lectures, conferences and my frequent chaotic discussions with family and friends.

While wandering though environments focused on sound and real-time 3D for more than 30 years, I have encountered (and still encounter) a great many problems for which I have had to devise various solutions, some of which are better than others.

Before we begin, we should take a moment to review the modern state of the industry with a few definitions, observations and important historical facts.

What is a sound effect?

Agreeing upon a definition is not easy, since the concept of sound effect can have several meanings. The first concepts that spring to mind are the very popular sound effects used in radio and movies to associate a certain motion, image, action, dialogue or commentary with a specific sound: an opening door, a galloping horse, the waves of the ocean, rain, a steam locomotive entering the station, car tires screeching on tarmac and so on.

This is the most common meaning of the term. Next, many people think of natural phenomena: echoes, the Doppler effect[1], the rumbling of thunder, sound masking[2], etc.

Third, it is music. Classically, we have the notion of tempo (adagio, allegro, vivace, etc.), as well as effects created by specific instrumental techniques (glissando, tremolo, vibrato, trills, etc.).

Finally, we have the idea on which this book is based. Namely, the analog and digital musical sound effects used in studio and live environments. The following definition is in my opinion the most insightful:

Generic term describing a modification of the sound parameters of a signal. It refers to techniques widely used in modern music (jazz, blues, rock, pop, etc.) to modify the original sound of an electrical or electronic instrument. The most common effects are: reverb, echo (delay), phasing, distortion, flanger, compression. Most modern recordings use the possibilities of sound effects in some form, either by means of expanders and multi-effect pedals, or by using an array of independent pedals. (www.musicmot.com)

If we had formulated the question more precisely and specifically, like: "What is an audio effect?", we might have arrived at a different answer. Consider for instance the following two quotes:

Audio effects are analog or digital devices that are used to intentionally alter how a musical instrument or other audio source sounds. Effects can be subtle or extreme, and they can be used in live or recording situations. (blog.dubspot.com[3])

---

1 The Doppler effect is a frequency shift phenomenon that occurs with mechanical waves (see section 5.4.2).
2 The practice of playing a sound, usually neutral or soft, to cover up unwanted noises and improve the feel of the sound quality in a location.
3 Music production and DJ school based in New York, Los Angeles, and online.

Sound effects (or audio effects) are artificially created or enhanced sounds, or sound processes used to emphasize artistic or other content of films, television shows, live performance, animation, video games, music, or other media. (Wikipedia)

With this definition of sound effects, one aspect is often important. Natural sound phenomena always unfold in real time, and of course the same is true for live performances. However, in studios, during postproduction, the notion of time suddenly becomes completely relative. Recorded audio signals do not fade over time.

The constraints are completely different, and real-time effects no longer have to play by the same rules.

When you are playing live music, you cannot go backwards in time; what is done is done. But in a studio, although listening and editing in real time can be more comfortable, you can always start again. Recorded tracks are not volatile, and even while you are still recording you can repeat anything you need to.

Let us now take a different approach, and consider the evolution of sound effects through history.

Our tale begins at the turn of the 20th Century[4].

Of course, there had been plenty of interest in sound effects around the world before then, but this had been largely restricted to musical instruments and singing (percussion, xylophones, flutes, church organs, etc.) without any connection to advanced technology based on electricity, electromechanics or electronics, which did not yet exist.

As we mentioned earlier, the first sound effects were widely used on the radio to provide an acoustic background for serial programs, which were very popular in the early 1920s.

Once movies progressed from silent pictures to pictures with sound, the movie industry also started to play an important role in sound effects.

By the early 1930s, most studios could mix manual sound effects and prerecorded sounds.

---

4 Author's note: The various effects mentioned here will be revisited, presented and explained in more detail in the chapters of this book.

The first magnetic recorder (called the *Telegraphone*) was invented by V. Poulsen[5], who used a wire and later a steel ribbon. F. Pfleumer[6] improved the design in 1928 by introducing a much more reliable recording medium, magnetic tape (paper tape or acetate and iron oxide), which gained prominence in the 1940s. This marked the beginning of the first usable recording devices.

Figure I.1. *A replica of the "Telegraphone", 1915–1918 (Gaylor Ewing collection) (source: museumofmagneticsoundrecording.org)*

One of the first effects used in musical recordings was reverb, which was used as early as 1930, followed by the appearance of the tremolo effect, based on an invention by D. Leslie[7], the creator of the famous speaker bearing his name.

The advent of analog electronics, first in the form of electronic tubes, then transistors, shook the musical world to its foundations, and sound effects began to take early recording studios by storm.

---

5 Valdemar Poulsen, 1869–1942, Danish engineer, inventor of the first magnetic recording device.
6 Fritz Pfleumer, 1848–1934, Austrian engineer, inventor of magnetic tape for sound recordings.
7 Donald Leslie, 1911–2004, inventor of the Leslie speaker, a mechanical and electronic amplification device originally designed for Hammond organs that creates a sound playback environment based on the Doppler effect.

The invention of magnetic tape recorders at around the same period created significant added value, opening new doors and new horizons for technicians and musicians, notably including electronic music and concrete music.

One easily overlooked sound effect is signal fading (with a potentiometer/fader), and mixing with other signals. Mixing also began to be used by radio stations around the early 1930s.

Once the first amplification systems began to hit the stage, mixing consoles were no longer limited to simply fading and mixing signals, but could now amplify them too.

In the mid-1940s, in his studio workshop in Hollywood, Les Paul[8] invented the delay and flanger[9] effects. His next project was to modify one of the first Ampex[10] tape recorders by adding extra recording heads, turning it into a multitrack recorder.

Figure I.2. *Brochure presenting the first Ampex tape recorders*

In the early 1950s, the first tone control circuits and filters began to appear. The well-known *Baxandall* tone control system, named after its inventor, P. Baxandall[11], could correct two or three bands (bass, mid and treble). It was soon followed by the first *equalizer*, marketed in 1951 by Pultec.

---

8 Lester William Polsfuss ("Les Paul"), 1915–2009, American guitarist and inventor.
9 This effect can be heard in the 1945 track "Mamie's Boogie".
10 American company, manufacturer of electronic products including the first studio tape recorders. Ampex is an acronym for Alexander M. Poniatoff Excellence. A. M. Poniatoff was the founder of the company.
11 Peter Baxandall, 1921–1995, British engineer and audio electronics specialist.

Figure I.3. *View of a mixing console in the 1960s*

Subsequent advancements introduced dynamic signal processing: *compression*, *limitation* and *expansion*.

As early as 1959, B. Putnam[12] came up with the idea of a modular mixing console with filters on each track.

During the same period, the distortion effect made its first appearance in amplifiers based on mechanisms ranging from holes in the membrane of the (electroacoustic) speaker to saturation of the amplifiers themselves.

Phasing would not be embraced by the music industry until 1975, despite having been known since 1940. *Pitch-shifting* was created at around the same time.

By 1977, the first digital audio workstations[13] had begun to arrive on the market. At this point, they were still extremely primitive, due to the low processing

---

12 Bill Putnam, 1920–1989, American engineer and audio specialist, composer, producer and entrepreneur. He is often described as "the father of modern recording".
13 The American company Soundstream was one of the first companies to market a DAW. It was based on a DEC PDP-11/60 mini computer.

capabilities of contemporary computers, but they have never stopped improving ever since.

Our story ends here, in the early 1980s. By this time, most sound effects had arguably already been invented, with the exception of *autotune* in 1997. Since then, the only real change has been recent integration in the form of software components (*plugins*) since the beginning of the 21st Century.

In conclusion, note that many of the sound effects presented throughout this book are currently extremely popular and have multiple areas of application. No doubt many of the audio enthusiasts reading this book will have already worked with some sound effects, and, if so, I hope that I will be able to provide some insight. For those of you who have not experienced this, I hope that you will enjoy discovering new ideas. Use them wisely, to enhance your sound recordings, mixes, your instrumental technique or even just your musical knowledge.

One final remark is that the many tables provided in the chapters of this book that group effects together by type do not attempt to establish an exhaustive index of everything that you might possibly encounter in the world of musical effects. New equipment is released every day, just as older models are gradually discontinued by their manufacturers and distributors over time. Electronic hardware and software also evolve at a breakneck pace, rendering them highly volatile and quickly obsolete.

1

Notes on the Theory of Sound

This chapter presents concepts that are indispensable to understanding and studying the phenomena associated with sound. In this study, readers can find the details required to discuss the variety of ways that exist to process sounds.

Mathematical equations are deliberately presented in their simplest and purest forms, without going into detail and avoiding any mathematical proofs. However, the standards of scientific rigor to which all physical sciences are committed have not been lost.

1.1. Basic concepts

We shall begin by describing the nature of a sound, followed by a few of its characteristics, before turning to the question of how our ears work. This leads into an analysis of the *typology of sounds*, *spectral analysis* and *timbre*.

To conclude, we will present the fundamental aspects of the propagation of sound, and we will consider a few common phenomena to be explored in more detail in subsequent chapters.

1.1.1. *What is sound?*

Although this question might seem simple, the answer is by no means easy. There are two ways to approach this topic: from a purely scientific perspective, working from the laws of physics, or alternatively by thinking about how our senses allow us to perceive sound.

Physicists view sound as a *mechanical wave* that propagates as a perturbation within an elastic medium or object. In other words, they view sound as the forward- and-backward motion (mechanical oscillation) of particles around their resting position. Electromagnetic waves, on the other hand, propagate as energy in the form of an electric field coupled to a magnetic field.

Figure 1.1. *A simple example of a mechanical wave. Here, the wave is created on the surface of the water after a stone is thrown in*

Many people will find it easier to define sounds as auditory sensations.

Sounds are produced by vibrating objects. These objects are sources, and the environment in which the sound is emitted carries the sound to our ears. When a sound reaches our ears, our brains allow us to perceive it, become aware of it and interpret it.

Most of the objects around us can produce sounds when they interact with shocks, friction, airflows or deformations. Who has not entertained themselves by twanging a plastic ruler at the edge of a table?

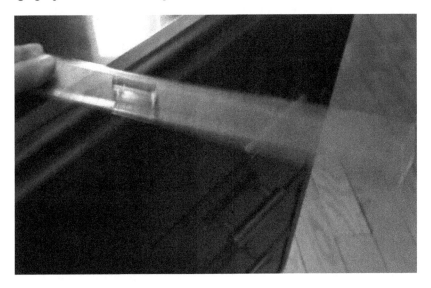

Figure 1.2. *Vibrating ruler at the edge of a table*

Sound cannot propagate in a vacuum, since there is no medium to convey the vibration.

Like all physical phenomena, sounds can be characterized. We can use parameters such as the *intensity*, the *pitch* and the *timbre* to define and distinguish different sounds. Other parameters are possible too – people interpret sounds as they hear them. This involves subjective phenomena, so-called *psychoacoustic* phenomena, which depend on the physiology, culture and ethnicity of the individuals receiving the sound message. This approach to decrypting the process of sound characterization is highly complex. We will quickly encounter questions to which science has not yet been able to provide comprehensive answers.

Over the centuries, philosophers have often wondered: does a sound exist if no one is there to hear it?

The field of science that studies the physics of sound phenomena is known as *acoustics*. Some of its goals include characterizing the *audibility* of sounds, defining the means of transportation and transformation of sounds and determining the deformations that a given sound can undergo. It is often an extremely theoretical field. Acoustics provides working foundations, but quickly reaches its limits when sounds come together to form music. Music is not a science, and appeals to many parameters that are not necessarily measurable, and which are intricately intertwined in extremely complex ways.

Before we continue, we shall review a few concepts that are required to understand the foundations and the nature of sound.

1.1.2. *Intensity*

This parameter characterizes the strength of a sound, determining whether the sound seems loud or quiet. The term *loudness* is also used.

When a sound is emitted, the sound waves deform the fluid medium (usually air). This deformation creates a change or local perturbation in the pressure around the source of the sound. This perturbation travels through the surrounding material(s) at a speed (*speed of sound* or *celerity*) that depends on the nature of the elements in the medium, as well as their states and thermodynamic properties. The materials through which the perturbation travels – fluids or alternatively flexible or rigid bodies – each have a certain *elasticity*, which usually allows them to recover their original states once the sound wave(s) have passed through them. Permanent deformation or destruction can occur if the sound pressure generated by the wave is greater than the *elastic limits* of these materials. This scenario is uncommon, since the stresses involved in sound propagation are relatively low.

The pressure variation mentioned above, usually observed in air, is known as the instantaneous acoustic pressure, measured in *pascals*. This acoustic pressure induces an acoustic energy. These two parameters are both described by the term *acoustic sound pressure level*, also simply called the *sound level*.

The sound pressure ranges over many different scales. Its units are *pascals* (1 Pa = 1 N/m^2), denoted by the symbol Pa. The normal atmospheric pressure at sea level is defined as 1.013×10^5 or 1,013 hPa (1 hPa = 100 Pa). When working with the scale of the acoustic sound pressure, we normalize by a reference pressure value close to the average absolute intensity threshold of the human auditory system between 1,000 and 4,000 Hz (hertz[1]), namely 2×10^{-5} Pa. This corresponds to a

---

1 Unit measuring the frequency of a phenomenon that has a period of 1 s, see section 1.1.3 of this chapter.

power of 10^{-12} W (watt^2)/m^2. However, these numbers are not easy to manipulate and are not very "intuitive", not to mention the fact that the ratio between the lowest audible sounds and extremely loud sounds is 1/10,000,000.

The sensation of sound is first and foremost a physiological phenomenon, and one useful approach is to quantify the sensation of excitation in a way that takes into account the acoustic range of the human ear, which works according to a logarithmic scale. The inventor of the telephone, Graham Bell[3], originally defined the *bel* as a base unit. Today, we often use the unit of one-tenth of a bel, the *decibel* (dB), as a simple way to quantify sound phenomena.

$$N_{dB} = 20 log_{10} \frac{P}{P_0} = 20 log_{10} \frac{P}{10^{-5}}$$

or

$$N_{dB} = 10 log_{10} \frac{W}{W_0} = 10 log_{10} \frac{W}{1 \times 10^{-12}}$$

where:

- N: sound level in decibels;
- P: measured acoustic pressure;
- P_0: reference acoustic pressure (0 dB);
- W: measured power;
- W_0: reference power.

The logarithmic growth of the decibel scale means that the sound intensity doubles every 3 dB.

---

2 Unit of power. One watt is the power of a system that receives a constant energy flux of 1 J/s.

3 Alexander Graham Bell, March 3 1847–February 8 1922, American inventor originally from Great Britain, born in Edinburgh. He invented the telephone in 1876 and founded the telephone company bearing his name. Creator of the "National Geographic Society".

Type of audio signal	Effects	Sound level (dBA)
Rocket take-off		180
Turbojet engine, airplane take-off		140
Rifle shot, engine on a test bench		130
Formula 1, jackhammer	Pain threshold	120
Rock band, metal workshop		110
Train passing by, circular saw, night club		100
Portable music player at maximum volume, sander, shouting	Hearing risk threshold	90
Radio at maximum volume, machine tools		80
Noisy restaurant, office with typewriters	Office work	70
Lively conversation, street, public place		60
Quiet conversation, large quiet office	Intellectual work requiring high concentration	50
Quiet apartment, quiet office		40
Walk through the forest		30
Peaceful countryside, whispering		20
Recording studio		10
Silence	Audibility threshold	0

Table 1.1. *Table of sound intensities*

When measuring the sound level with a sonometer, our unit of choice is the *decibel A* or *dB(A)*. As we will see in section 1.2.2, the sensitivity of the human ear varies as a function of the frequency of the sound signal. To compensate for this physiological behavior of our ears, the sound levels of each frequency component of a sound wave are weighted and summed to give an overall measurement. The units of dB(A) are adjusted to reflect this weighting.

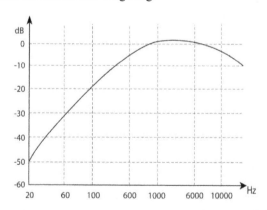

Figure 1.3. *Psophometric curve (weighted curve) db(A)*

1.1.3. Sound pitch

The pitch of a sound is characterized by its *frequency*, i.e. the number of oscillations per second of the molecules in the traversed medium (usually air) around their resting position when a sound wave passes through them.

The frequency is measured in units of *hertz*, denoted by the symbol Hz, and its multiples: kilohertz (kHz), megahertz (MHz), etc. The range of audible frequencies for humans is 20–20,000 Hz (20 kHz). This range varies from individual to individual, and also changes with age.

The speed at which a sound wave travels, also known as its celerity, is 343 m/s through air at a temperature of 20 °C. This value changes depending on the nature of the object or the medium through which the wave is traveling, as well as the pressure and the temperature. For instance, it is equal to 331 m/s through air at 0 °C. The speed of sound is higher in liquid and solid objects (~1,400 m/s in water and ~5,000 m/s in steel).

Having defined the frequency, we should take the opportunity to define a few other parameters: the *wavelength*, the *period* and the *amplitude*.

The wavelength is the distance between two consecutive maxima of a sound wave. This defines the separation between two consecutive periods of a periodic wave (see Figure 1.4), and so is also equal to the distance traveled by the wave during one period.

The period is the time in seconds taken to complete one full oscillation (one cycle). It is the inverse of the frequency and vice versa:

$$T = \frac{1}{f} \text{ and } f = \frac{1}{T}$$

where:

– T: period in seconds;

– f: frequency in hertz.

A frequency of 1,000 Hz (1 kHz) corresponds to a period of 0.001 s or 1 ms.

The amplitude defines the sound intensity. As we saw above, this characterizes the variation in pressure.

The wavelength, the period, the frequency and the speed of sound satisfy the relation:

$$\lambda = cT = \frac{c}{f}$$

where:

– λ: wavelength in meters;

– c: speed of sound in m/s;

– T: period in seconds;

– f: frequency in hertz.

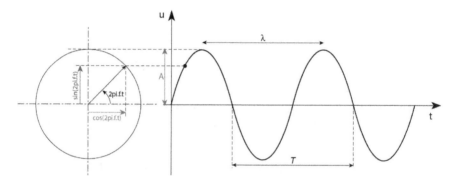

Figure 1.4. *Representation of a sound wave and its parameters over time*

1.1.4. *Approaching the concept of timbre*

Timbre is what allows our ears to distinguish and recognize different sounds, whether everyday noises, musical instrument or people's voices.

It is closely related to the shape of the sound wave. Timbre is a complex notion that we still struggle to fully explain today, since it involves concepts relating to the act of hearing, our ability to judge sounds, our auditory memory and perception.

Before we can begin to discuss the notion of timbre more precisely, we will need to study several other sound-related parameters, including the physiological mechanisms of our ears, the typology of sounds, the spectrum, transient phenomena and the nature of sound-emitting source(s).

Claude Elwood Shannon[4], a renowned mathematician, once stated: "Timbre is what makes sounds sing to our ears".

1.2. The ears

Hearing is the second of our five senses. It relies in part on our auditory system, whose primary external organ is our ears.

In this chapter, we will find out precisely how our ears work, what makes them interesting and we will analyze how they operate within sound-based environments.

1.2.1. *How our ears work*

The ear can be divided into three parts: the outer ear, the middle ear and the inner ear.

The outer ear is the part of the system that captures sounds. This system serves the roles of amplification and protection. It is separated from the middle ear by a thin, flexible membrane, the eardrum, which deforms under the effect of sound waves.

The outer ear consists of the *auricle* and the *ear canal*, the latter of which is approximately 2.5 cm long. The ear canal carries sound vibrations to the *eardrum*, amplifying frequencies between 1,500 and 7,000 Hz by 10–15 dB. These are the most useful frequencies to us, notably including speech.

The auricle also plays a role in locating the source of sounds in space for frequencies between 2,000 and 7,000 Hz.

---

4 Claude Elwood Shannon, 1916–2001, American electrical engineer, mathematician and cryptographer, known as the father of the digital transmission of information. He created the field of "mathematical information theory".

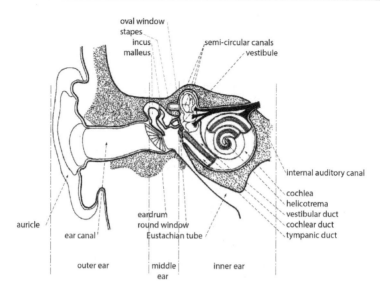

Figure 1.5. *The outer, middle and inner ear*

The *middle ear* is a cavity filled with air. It is connected to the *pharynx* by the *eustachian tube*, which opens when we swallow in order to equalize the pressure on either side of the eardrum. This cavity also contains a *series of ossicles* (small bones): the *malleus*, attached to the eardrum, the *incus* and the *stapes*. The stapes acts as an interface between the air-filled medium of the middle ear and the liquid medium of the inner ear. It rests against the *oval window*, which acts as the boundary to the *inner ear*.

Together, the ossicles form a lever that increases the pressure and thus the amplitude of sound waves. The surface area of the eardrum is around 15 times larger than the oval window, which creates an increase of 20 dB. The middle ear acts as a pressure amplifier.

When excessive sounds louder than 80 dB are detected, the *stapedius* (stapes muscle) contracts to reduce the vibration of the ossicles (*acoustic reflex*), attenuating the transmission of sound waves to the inner ear. This protection mechanism reduces the sound signal by 40 dB.

It should be noted that the stapedius develops fatigue over time, and so cannot provide long-term protection. Additionally, it only activates at low frequencies of less than 1 kHz, and the contraction occurs too slowly to protect from sudden noises like explosions, since the "reflex latency" (physiological reaction time associated with the information processing sequence of humans) is around 30 ms.

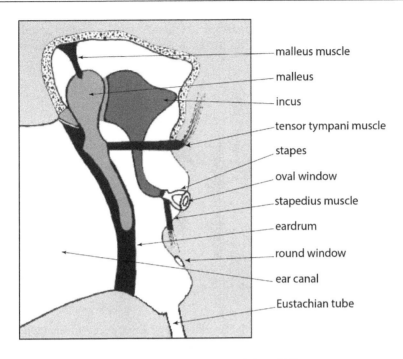

Figure 1.6. *Detailed diagram of the middle ear*

The inner ear consists of two sensory organs: the *vestibule*, a balance organ, and the *cochlea*, a hearing organ.

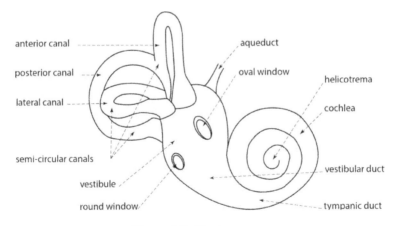

Figure 1.7. *The inner ear*

The cochlea is a hollow bone shaped like a snail with a 2.5-turn spiral. It has three canals: the *vestibular duct*, the *tympanic duct* and, in the center, the *cochlear duct*.

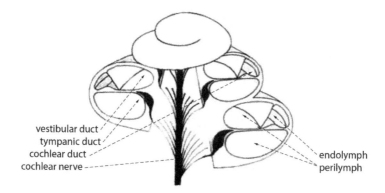

Figure 1.8. *The cochlea*

The vibrations applied to the oval window by the stapes are transmitted via the oval window into a sodium-rich liquid called the *perilymph* inside the vestibular duct, which is connected to the *apex* (tip) of the cochlea, in a region called the *helicotrema*. The round window in the tympanic duct compensates for the expansion of the liquid.

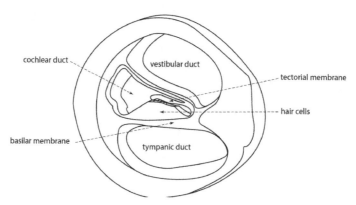

Figure 1.9. *Cross-section of the cochlea*

The cochlear duct is closed and filled with potassium-rich *endolymph*. It is separated from the tympanic duct by the *basilar membrane*. The sensory organ carrying the receptors, called the *organ of Corti*, is located on the internal face of this membrane. The receptors are *hair cells*. The inner hair cells are arranged in a single row, and the outer cells are arranged in three rows. The tips of the hairs (or *stereocilia*) of the outer cells are anchored to the *tectorial membrane*. Both inner and outer cells are connected to the fibers of the *auditory nerve*. The motion of the liquids contained in the ducts induces deformations in the basilar membrane, which tilts and twists the hairs connected to the tectorial membrane, which itself remains fixed. This tilting and twisting encodes the sound vibrations into ionic motion that polarizes or depolarizes the membrane of each cell.

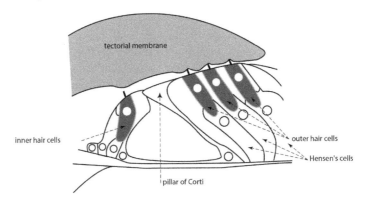

Figure 1.10. *Detailed diagram of the organ of Corti*

The outer hair cells are in fact selective amplifiers, and the inner cells are the actual sensory cells. A *tonotopy* (representation of the auditory spectrum) is arranged along the cochlear duct. In other words, a characteristic resonance frequency can be determined at each point along the duct. The low-frequency resonators (low-pitched) are located near the apex, and the high-frequency resonators (high-pitched) are located at the base of the duct.

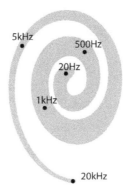

Figure 1.11. *Tonotopy of the cochlear duct and frequency distribution*

As well as being channeled through the air, sound vibrations are transmitted by means of another mechanism, known as *bone conduction*. Sound messages are directed to the inner ear via the bones in the skull. A much larger quantity of energy is required to produce a given stimulus via this path than is needed to propagate sound through the air. Indeed, the relative attenuation between the two paths has been measured as 30–60 dB, depending on the frequency.

1.2.2. Fletcher–Munson curves

The sensitivity of our ears is not linear with respect to the sound pressure. In other words, the perceived volume of a sound with a given intensity can seem higher or lower depending on the frequency of the signal. The *Fletcher–Munson curves* demonstrate this phenomenon. Below a certain frequency-dependent power threshold, sounds are imperceptible. This defines our threshold of hearing. The same is true for high-power sound messages, which become unbearable after a certain point, defining the threshold of pain. Fletcher established the curve relating the frequency on the *x*-axis to the power (sound pressure) on the *y*-axis. When listening *binaurally* (with both ears), the curves show which sounds create identical sensations. This work was standardized in 1961 to define so-called *loudness contours* (*isosonic curves*).

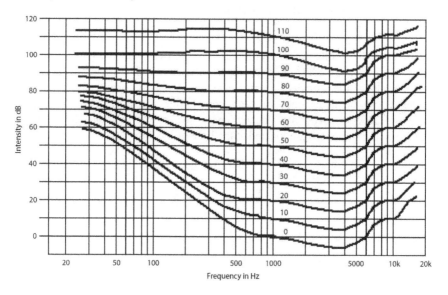

Figure 1.12. *Fletcher–Munson curves (isosonic curves)*

1.2.3. *Auditory spatial awareness*

Our ears are capable of pinpointing the source of a sound fairly accurately. This ability is based on several parameters.

In 1907, Lord Rayleigh[5] demonstrated the principles of *interaural level differences* and interaural time differences for the first time.

The *interaural time difference* (ITD) is a construction characterizing the time difference between the arrival of a sound wave at each of the two ears of a person. If the sound is coming from the front, this difference is zero.

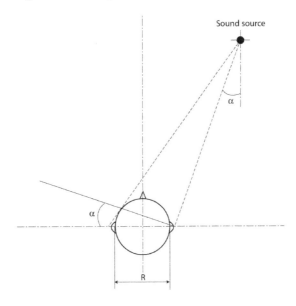

Figure 1.13. *Principle of sound localization by ITD*

The ITD can be approximated with the equation:

$$\Delta_t = \frac{R(\alpha + \sin\alpha)}{c}$$

---

5 John William Strutt Rayleigh, British physicist born at Langford Grove (Essex), 1842–1919. He was awarded the Nobel prize in physics in 1904 for the discovery of inert argon gas together with William Ramsay, and conducted extensive research into wave-related phenomena.

where:

- Δ_t: ITD in seconds;
- R: radius of the head in meters (8.75 cm by default);
- α: angle of incidence in radians;
- c: speed of sound (340 m/s).

For example, if a sound source is located at 30°, we find that:

$$\Delta_t = \frac{0.0875(0.523 + 0.5)}{340} = 2.63 \times 10^{-4}\text{s} = 263\ \mu s$$

The sound will reach the person's ears with a time difference of 263 μs. This difference can be thought of as a phase shift that is analyzed and interpreted by our brain in order to locate the position of the source. The maximum ITD is around 673 μs.

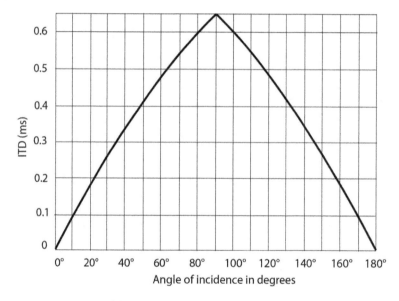

Figure 1.14. *ITD calculation chart*

This phenomenon is particularly distinctive at low frequencies below 1,500 Hz. At higher frequencies, the interaural level difference (ILD), the interaural intensity difference (IID) or the interaural pressure difference (IPD) is used instead.

If a sound is emitted by a source that is closer to one ear than the other, there will be difference in the sound intensity or acoustic pressure. Our auditory system uses this difference to locate the sound source.

Gary S. Kendall and C.A. Puddie Rodgers proposed a simple equation to calculate the ILD:

$$\Delta_l = 1 + \left(\frac{f}{1,000}\right)^{0.8} \times \sin\alpha$$

where:

- Δ_l: ILD;
- f: frequency in kHz;
- α: angle of incidence in radians.

But the ITD and ILD alone are not the only factors that allow us to discriminate between the positions of different sources. In some cases, the ITD and the ILD may be identical, even though the sources are located at different positions, as shown in Figure 1.15.

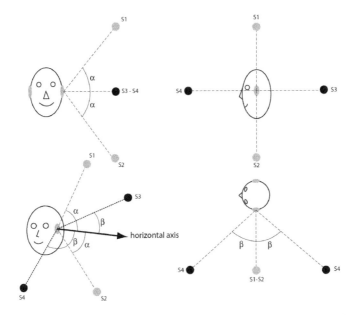

Figure 1.15. *Angular localization error in the horizontal and vertical directions. The sources S1 and S2 (vertical plane) have exactly the same ITD and ILD values as the sources S3 and S4 (horizontal plane)*

Another parameter is also used as a factor to locate the source of sounds based on diffraction caused by the morphology of the head. This eliminates the ambiguity created by the localization errors presented previously. Today, this factor is thought to be the most important factor, and is currently the subject of extensive research. The head-related transfer functions (*HRTF*) method represents the result of this work.

To understand how HRTF localization works, suppose that a source is placed directly to the right of an individual. His or her right eardrum will receive the sound message along a straight path. However, the sound waves that reach the left ear will need to follow a much more complex path around the head before ultimately hitting the left eardrum after multiple reflections and diffractions within the auricle and the ear canal.

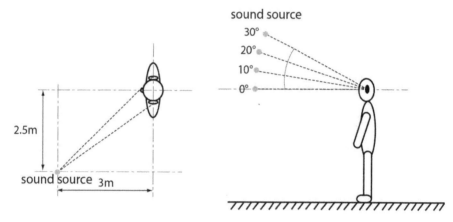

Figure 1.16. *Implementation of HTRF. The source is placed in front of the observer, and its elevation is varied from 0 to 30° relative to the horizontal plane through the observer's ears*

As it travels, the timbre of the sound wave, and therefore its spectrum, is modified. These modifications depend on the location of the source within three-dimensional space and are interpreted by the brain in order to determine this location. Note that HRTF localization is capable of determining vertical position, unlike the ITD and ILD.

Notes on the Theory of Sound   19

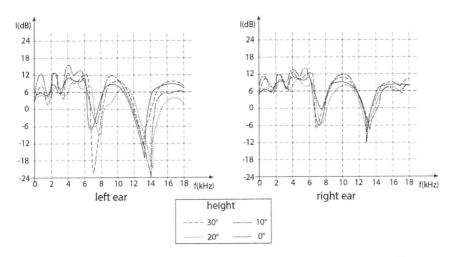

Figure 1.17. *The HRTF can distinguish between different heights, unlike the ITD and ILD methods. The curves change as a function of the height of the source relative to the ears of the observer*

The angular localization errors[6] are shown in Figure 1.18, determined by all of the above methods (ITD, ILD and HRTF).

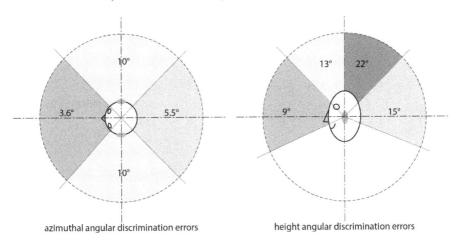

Figure 1.18. *Values of the angular perception errors in the horizontal and vertical directions*

---

6 Based on the measurements and experiments performed by Blauert.

It should be noted that the localization functions studied previously (ITD, ILD and HRTF) are based on binaural hearing (with both ears), but localization is also possible with monaural hearing. The auricle of the outer ear filters sound by means of reflections that depend on the angle of incidence of the source. These reflections introduce delays and timbre deformations, which can be used to determine the origin of the emitted sound.

Certain effects can impede or enhance the process of locating sounds in space. Two effects are particularly significant: the *"cocktail party"* effect, and the precedence effect, also known as the *Haas effect*.

The "cocktail party" effect occurs in noisy environments. We are capable of locating the source of a sound if we know where it is.

The Haas effect[7] reveals how we perceive reflected sounds compared to direct sounds. Our ears cannot distinguish between direct sounds and reflected (or reemitted) sounds if they are separated by less than 50 ms, even if the reflected source is louder[8] than the direct source by several dB. If the separation is longer, our ears no longer perceive the sources together, but instead as an echo. This phenomenon implies that our auditory system interprets the direction of an acoustic phenomenon as the direction of the first source that it perceives. This effect is also known as the law of the first wavefront.

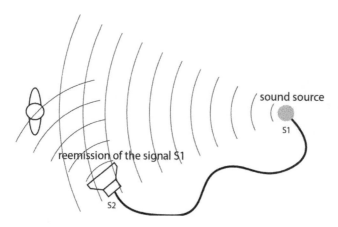

Figure 1.19. *Demonstration of the precedence effect or Haas effect. If the difference between the sources S1 and S2 is less than 50 ms, the observer cannot perceive a difference between them (no echo)*

---

[7] According to a study conducted by Helmut Haas in 1951.

[8] Depending on the separation (0 to 50 ms), this value ranges up to a maximum of 10 dB.

1.3. The typology of sounds

Like everything that surrounds us, sounds can be organized into a typology based on characteristics distinguishable by physics or hearing.

1.3.1. *Sounds and periods*

Pure sounds and constant frequencies rarely exist in our natural environment, except in certain circumstances. The sounds around us are typically complex. We can define a typology of sounds to distinguish them. The first thing to consider is the property of periodicity.

A sound is said to be *periodic* if its frequency does not change over time.

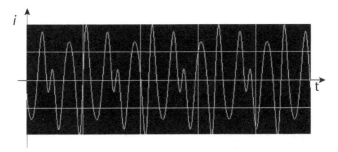

Figure 1.20. *Representation of a periodic sound*

A sound is said to be *aperiodic* if it is characterized by a large number of changes in frequency and amplitude over time. Most of the noises in our environment are aperiodic.

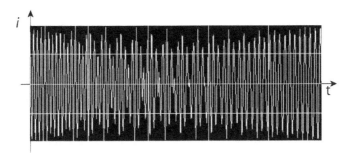

Figure 1.21. *Representation of an aperiodic sound*

White noise is an extreme example of an aperiodic sound. It uses the entire spectrum of audible frequencies.

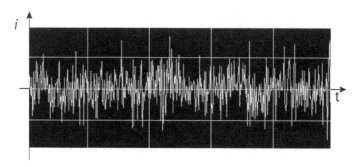

Figure 1.22. *Representation of white noise*

A sound with extremely short duration is called an *impulse*, whereas longer sounds are described as *continuous*.

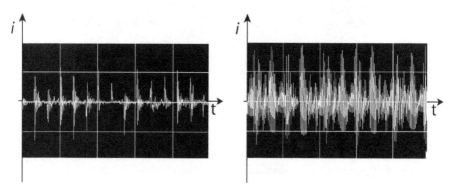

Figure 1.23. *Representations of impulses and a continuous sound*

1.3.2. *Simple sounds and complex sounds*

Sounds with sinusoidal waveforms are called *simple sounds*. All other sounds are *complex sounds*. Complex sounds are composed of two or more sinusoidal waves. Thus, complex sounds are combinations of multiple simple sounds.

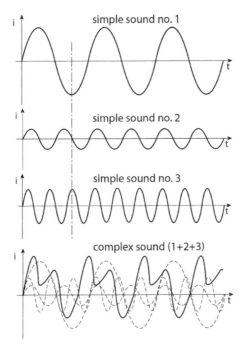

Figure 1.24. *Combination of multiple simple sounds*

To determine the simple waves that compose a *complex sound*, we can use a mathematical tool called a *Fourier transform*[9].

Fourier transforms allow us to decompose complex sounds into a series of simple sounds (sinusoidal waves). Complex sounds in the real world can contain dozens of sinusoidal components.

The amplitude of a complex sound at any given moment in time is given by the sum of the amplitudes of the simple sounds from which it is composed.

The frequency of a complex sound is equal to the lowest of the frequencies of the simple sounds from which it is made up of.

The frequency of a complex sound is called the *fundamental frequency* (F0). If a sound is not periodic, it implicitly does not have a fundamental frequency. Instead, we think of it as a noise.

---

9 Jean Baptiste Joseph Fourier, March 21 1768–May 16 1830, French mathematician and physicist.

1.4. Spectral analysis

Spectral sound analysis combines several analytical techniques for determining the characteristics of an audio signal. The results of spectral analysis are often presented graphically.

1.4.1. *The sound spectrum*

The spectrum of a sound is a representation of its amplitudes and frequencies. The representations that we used earlier always described each sound by a variation in amplitude along the vertical axis and a horizontal axis representing time.

The spectrum of a sound wave does not contain any information about time.

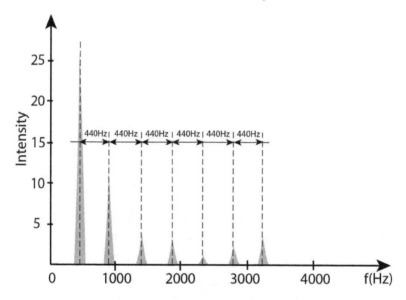

Figure 1.25. *Spectrum of a periodic complex sound*

The spectrum is often represented as a vertical bar graph (or line graph). Each bar shows the amplitude of a certain frequency. The lowest frequency is the fundamental frequency.

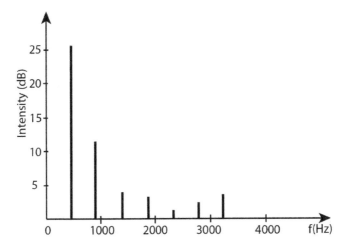

Figure 1.26. *Line spectrum of the sound in Figure 1.22 (intensity in dB)*

This type of graph is excellent for periodic sounds, since they contain a limited number of frequencies. We say that such sounds have a *discontinuous spectrum*, *line spectrum* or *comb spectrum*.

The other frequencies are called *harmonics* and are multiples of the fundamental frequency.

Designation	Symbol	Frequency
Fundamental frequency or first harmonic	F_0	440 Hz
Second harmonic	F_1	$2 \times F_0 = 2 \times 440 = 880$ Hz
Third harmonic	F_2	$3 \times F_0 = 3 \times 440 = 1,320$ Hz
Fourth harmonic	F_3	$4 \times F_0 = 4 \times 440 = 1,760$ Hz
Fifth harmonic	F_4	$5 \times F_0 = 5 \times 440 = 2,200$ Hz
Sixth harmonic	F_5	$6 \times F_0 = 6 \times 440 = 2,640$ Hz
Seventh harmonic	F_6	$7 \times F_0 = 7 \times 440 = 3,080$ Hz

Table 1.2. *The harmonics of (440 Hz)*

Sometimes we need to visualize the frequency distribution of the spectrum, in which case we use a graph showing the *spectral envelope* of the sound instead of a bar graph.

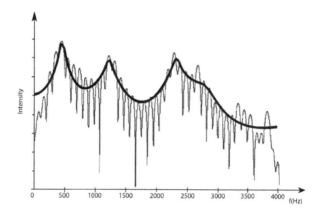

Figure 1.27. *Spectral envelope of a complex aperiodic sound*

This type of graph is excellent for aperiodic sounds with many frequencies. Sounds like this are said to have a *continuous spectrum*.

1.4.2. *Sonogram and spectrogram*

One of the disadvantages of the spectral representation, as we saw earlier, is that it does not include any information about time. However, when the signal is *pseudo-periodic* (a sequence of periodic signals over time: musical phrase, speech, etc.), it can be useful to take time into account.

Accordingly, scientists developed new types of graph with three parameters (frequency, intensity and time), known as *sonograms* or *spectrograms*.

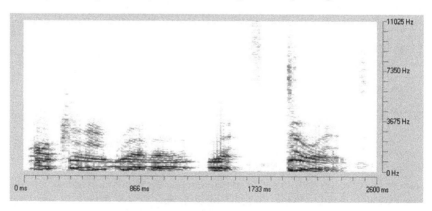

Figure 1.28. *Sonogram of a sound sequence*

A sonogram is a 2D graph where the sound intensity is defined by a scale of colors or shades of gray. A spectrogram is usually a 3D graph with time and frequency on the x- and z-axes, and the sound intensity on the y-axis.

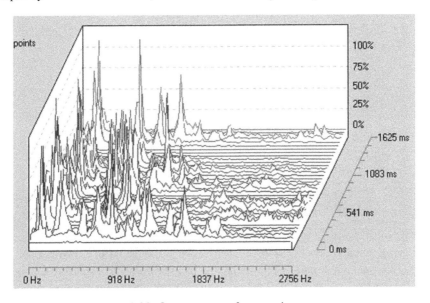

Figure 1.29. *Spectrogram of a sound sequence*

1.5. Timbre

The notion of timbre is too difficult to describe in terms of the fundamental frequency, the harmonics or the loudness (sound intensity). Timbre is a complex *psychoacoustic* concept. Sounds can be strident, dull, dry, warm and many more things.

1.5.1. *Transient phenomena*

Sound perception exists within a wider context. Each sound begins, stabilizes and then fades. Each single moment in the act of hearing a sound is a *transient phenomenon*. There are several types: *attack transients* and *release transients*, which describe the beginning and the end of sound phenomena. We can add other parameters to this list, which, like *vibrato* and *tremolo*, can either occur in a single phase (often when the sound has stabilized), or in every phase.

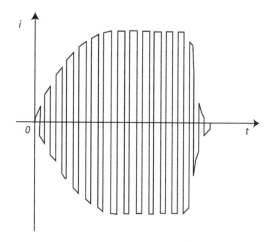

Figure 1.30. *The transients of the sound of a pipe organ*

The attack transient gives the sound its unique signature. For example, the attack transient is what allows us to distinguish between the sound of a clarinet and the sound of a flute. Its duration varies, typically ranging from 1 to 100 ms. It is thought that the human ear needs 40–50 ms to distinguish and recognize a sound. This type of transient is often complex. As well as its duration, we need to consider its *slope*, its instability, its spectral composition, the number of components and the order in which they occur, as well as many other factors.

The stabilization phase (or sustain) is often steady, although it can be influenced by the environment or the musician's technique.

The release transients often depend on environmental parameters (echo, reverb, damping, etc.) and technique-related parameters for musical instruments. It can range from very short to very long (sound accessories or natural or artificial echo phenomena), from 1 to 5,000 ms.

Together, the transient phenomena: attack, sustain and release are often called the *envelope* of the sound.

1.5.2. *Range*

Each sound source, especially musical instruments and voices, can emit sound over a certain range of frequencies. This range is also known as the *tessitura*. However, we must be careful to distinguish between two concepts: the *fundamental range* and the *spectral range*

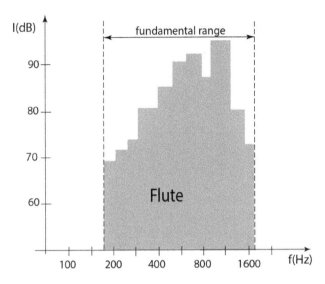

Figure 1.31. *Fundamental range of a flute*

The fundamental range (see Figure 1.31) only includes fundamental frequencies and harmonics. In an ideal setting, this is equivalent to the spectral range. However, reality often diverges from theory, which often attempts to describe the range of musical instruments simplistically as a set of fundamentals plus harmonics, with higher ranking harmonics having decreasing intensity.

The spectral range (see Figure 1.32) is the full spectrum of sounds that the source can produce.

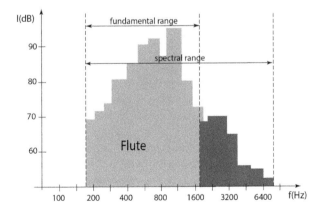

Figure 1.32. *Fundamental range of a flute compared to its spectral range*

1.5.3. Mass of musical objects[10]

This notion of mass was introduced by Pierre Schaeffer[11], who discussed it in his "Treatise on Musical Objects", which remains an important reference to this day. On the subject of concrete sounds, i.e. sounds that are not affiliated with any musical instrument that is culturally recognizable to a given observer, he writes:

> *But when we consider an arbitrary concrete sound (for example, produced by a membrane, a metal sheet, a rod...), we see that, unlike traditional sounds that have a clearly identifiable pitch, it has a certain* mass *located somewhere within its range, approximately characterizable by the intervals that it occupies, which are relatively easy to decipher. For example, it might contain several sounds with slowly changing pitch, which are dominated or surrounded by an aggregate of partials that are also gradually changing, the entirety of which can be approximately localized within a certain pitch interval. Our ears can quickly identify the most salient features and components, with some practice; these sounds then become as familiar to us as traditional harmonic sounds: they have a characteristic* mass.

His comments reveal the extent of the complexity involved in attempts to define the timbre. The concept of timbre extends far beyond the physics of the phenomenon into the realm of psychoacoustics. Schaeffer continues:

> *At this point, we should stop to note that, since musicians care about things like: a note with good timbre, good or poor timbre, etc., they are distinguishing between two separate notions of timbre: one that relates to the instrument, indicating the provenance of the sound by the simple act of hearing, and another relating to each of the objects created by the instrument, involving an appreciation of the musical*

---

10 The concept of object was particularly important to Pierre Schaeffer: "A sound object refers to the studied signal, viewed within the context of perception. Instead of hearing events via sounds, we hear sounds as events."

11 Pierre Schaeffer, 1910–1995, founder of the research department of the ORTF (Office de Radiodiffusion-Télévision Française), researcher in the fields of audiovisual communication and music. The inventor of musique concrète (concrete music), he created the GRM (Groupe de Recherche Musicale) in 1951, as well as being a composer (Variations sur une flûte mexicaine, Suite pour 14 instruments, Symphonie pour un homme seule, Toute la Lyre, Orphée 51, Masquerage, and others), and author of the monumental and prophetic work "Traité des Objets Musicaux" ("Treatise on Musical Objects").

effects contained in the objects themselves, which are desirable in the context of musical interpretation and musical activities. We can go even further by speaking of the timbre *of one single component of the object: the timbre of the attack, which is distinct from its stiffness.*

However, given this definition, the timbre of an object is nothing other than the form and substance of its sound, its complete specification among the range of sounds that a given instrument can create, up to any admissible artistic variation. Associating the concept of timbre with the object therefore cannot help us describe the object itself any further, since it simply postpones the analysis of the subtleties of our qualified perceptions of the sound.

1.5.4. Classification of sounds

Based on the observations relating to the concept of timbre presented above, we can define a classification of different types of sound message.

Category	Composition	Spectrum	Example
Pure sound	Tonal sound without harmonics	Filiform	Sinusoidal sound (BC generator, synth VCO, etc.)
Tonal sound	Sound with an identifiable pitch	A band	Note played by an instrument (piano, harpsichord, violin, etc.)
Tonal group	Group of multiple tonal sounds	Multiple bands	Chord played by an instrument (piano, harpsichord, organ chord, etc.)
Nodal group	Aggregate sound without an identifiable pitch	Multiple bands	Set of percussions (multiple cymbals together)
Nodal sound	Set of multiple nodal groups	A band	Percussive sound (cymbal)
White noise	Group containing every pitch	Full spectrum	White noise generator
Complex sound	Mixed group containing tonal sounds, tonal groups, nodal sounds, nodal groups.	Complex shape	Natural sound (bell, gong, metal sheet, etc.)

Table 1.3. *Classification of sounds*

1.6. Sound propagation

Sound waves propagate through their surrounding media by means of specific phenomena that result in specific behaviors. We will study the most important principles.

1.6.1. *Dispersion*

A sound wave emitted by a point source disperses as a set of concentric spheres.

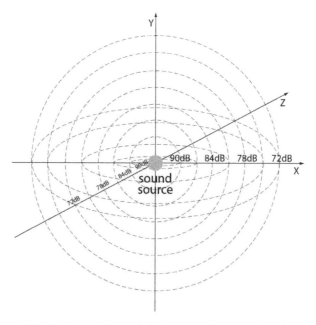

Figure 1.33. *Spherical dispersion of a wave from a point source. The sound pressure level decreases by 6 dB whenever the distance doubles*

Sound propagates through gaseous media like air, which is the most common transporting medium, in the form of alternating compressed and dilated layers. The set of wavefronts vibrate in phase (with each other).

In order to describe certain phenomena, scientists introduced the abstract notion of plane waves, which do not actually exist in reality. Plane waves are just sections of spherical waves. When the source is sufficiently far away from the point of audition, spherical waves have a large radius of curvature, and so may be approximated by plane waves.

1.6.2. Interference

Figure 1.34. *Interference between two waves on the surface of a liquid*

When two sound waves meet, they overlap, and their intersection creates *constructive* or *destructive interference*.

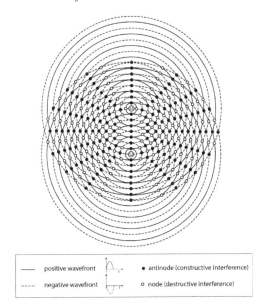

Figure 1.35. *Interference between two identical waves (same frequency and amplitude). The sources S1 and S2 emit sound waves that overlap, creating interference (nodes and antinodes)*

If you strike a tuning fork it and then rotate it near your ear, you will notice that it sounds louder or softer depending on the angle of rotation. This simple experiment demonstrates the creation of constructive and destructive interference.

Figure 1.36. *Interference created by a tuning fork*

When the sound waves emitted by each branch of the tuning fork are in phase, they add together to create constructive interference. But if they are perfectly out of phase, they generate destructive interference.

REMARK: There is an intermediate region between constructive interference (perfect sum of both signals) and destructive interference (perfect cancellation of both signals) in which the amplitude of the signal varies between its maximum and minimum values.

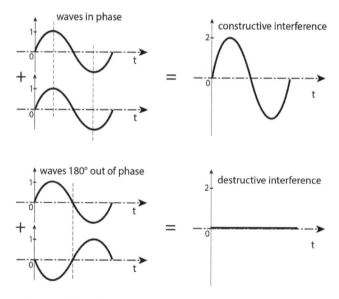

Figure 1.37. *Constructive and destructive interference*

Constructive and destructive interference[12]

With two identical sound waves, the shape of the interference pattern is easy to determine. In the case of non-identical waves, it can be determined by the *principle of superposition*: when two sound waves occupy the same space at the same time, the total perturbation is simply the sum of each of the two individual perturbations at each point in space and time.

1.6.3. Diffraction

When a sound wave encounters an obstacle, it goes around it. The edge of the obstacle becomes the center of a new wave, called a *secondary wave* or a *diffracted wave*.

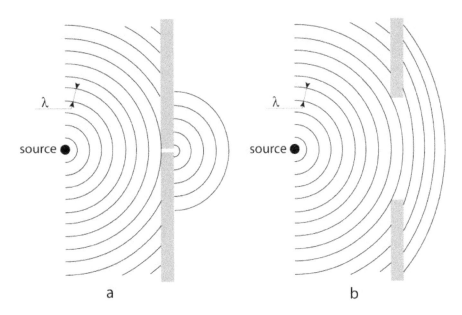

Figure 1.38. *Diffraction of a wave through an opening. The opening in a) is small relative to the wavelength λ, so a new point source is created. The opening in b) is larger than λ, so there is practically no diffraction*

---

12 We will return to the concepts of interference, phase and antiphase in section 4.1 when discussing equalization and filtering effects.

Diffraction is what allows us to hear sounds emitted by sources located behind obstacles.

Waves do not propagate in straight lines. Once a wave has moved past the obstacle, it again propagates in every direction. This phenomenon was demonstrated by the physicist C. Huygens[13], who also gave us *Huygens' principle*.

> "Each point on a wavefront may be viewed as the point source of a wave propagating in the same direction as the original wave. The next wave front may be obtained by summing all of these new waves".

In simpler terms, a wave can be thought of as a sum of elementary waves vibrating in phase that are moving in the same direction as the original wave.

When the obstacle placed in front of the sound-emitting source is smaller than the wavelength, it does not exist from the perspective of the wave – the sound wave propagates around it as if it were "invisible". The greater the wavelength-to-width ratio of the obstacle (λ/l), the greater the effect of diffraction. When the ratio is large, an observer located behind the obstacle does not perceive any difference compared to when the obstacle is removed.

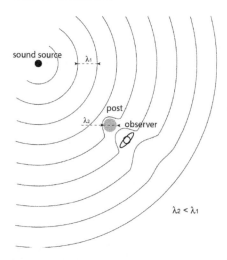

Figure 1.39. *Circumventing an obstacle smaller than the wavelength*

---

13 Christian Huygens, born at La Haye, Holland, April 14 1629–July 8 1695, mathematician, physicist and astronomer, author of "De ratiociniis in ludo aleae" in 1657 and "Horologium oscillatorium" in 1673. He discovered Saturn's ring, the rotation of Mars and the Orion nebula. He researched and published many papers on the wave theory of light, which enabled him to explain reflection and refraction, among other things.

If the width of the obstacle is large, part of the sound wave is reflected, and the edges of the obstacle become secondary sources of propagation.

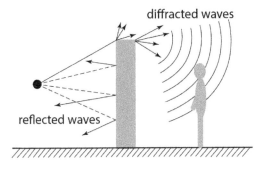

Figure 1.40. *Emission of secondary waves at the edges of an obstacle*

1.6.4. Reflection

When a sound wave encounters an obstacle that it cannot circumvent because the width of the obstacle is greater than the wavelength of the wave, the wave is reflected. The *reflection* of sound waves follows the appropriately named *law of reflection* (Descartes' law[14]). This law states that the new direction of the sound wave after hitting a point on the surface is equal to the angle of the wave with the normal at this point at the moment of impact; in other words, the *angle of incidence* is equal to the *angle of reflection*.

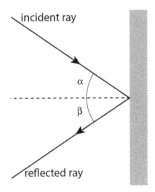

Figure 1.41. *Angles of incidence and reflection of a sound wave (α = β)*

---

14 René Descartes, born in France, Touraine, 1596–1650, philosopher, physicist and mathematician, author of the famous "Discours de la méthode", as well as his "Traité du Monde et de la Lumière", "la Dioptrique", "Météores" and so on.

Note that the reflected waves can interfere with the incident waves to create constructive or destructive interference. Within a thin layer very close to the surface hit by the wave, the sound intensity is increased by the sum of the incident wave and the reflected wave. In this layer, the acoustic sound pressure is doubled, which increases the intensity by around 6 dB.

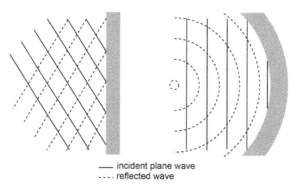

—— incident plane wave
---- reflected wave

Figure 1.42. *Reflection of a sound wave on a flat surface and a concave surface*

If the obstacle reflecting the wave has a concave shape, a *focusing* phenomenon occurs. In the opposite case, when the obstacle has a convex shape, a *scattering* phenomenon occurs.

Wavefronts hitting a wall created reflected waves that act as if they were generated by an "image" of the sound source located the same distance on the other side of the wall.

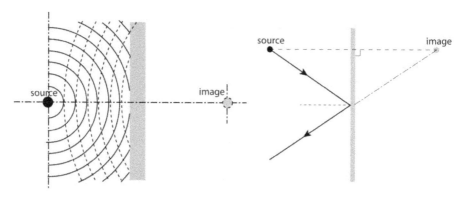

Figure 1.43. *Reflection of a sound wave on a wall and image of its source*

When a sound wave strikes a hard surface, there is no phase change in the reflected wave. We say that the surface has an *acoustic impedance* greater than air. But when a sound wave encounters an obstacle with a lower impedance, there is a phase inversion (for example when a sound transmitted by a solid meets the air).

1.6.5. Reverberation (reverb)

This phenomenon will be discussed at great detail in Chapter 8, which presents time effects.

1.6.6. Absorption

When a sound wave hits an obstacle, it is reflected, but loses some of its energy. This energy is absorbed by the material. The ability of a material to capture or absorb sound energy is described by its *absorption coefficient*, often denoted as α.

$$\alpha = 1 - |r|^2$$

where:

– α: absorption coefficient (recall that $\alpha = 1$: perfectly absorbent, $\alpha = 0$: perfectly reflective);

– r: reflection factor.

Hard materials such as marble, cement and plaster have absorption coefficients ranging from 0.01 to 0.05, whereas porous or fibrous materials such as carpet, felt or glass wool have coefficients between 0.2 and 0.4 (see the table of absorption coefficients established by Sabine in section 8.1.1).

1.6.7. Refraction

When a sound wave passes from one environment to a different environment, it changes speed (*discontinuous celerity*) and direction. This generates both a reflected wave and a *refracted wave* (especially in the case of thin materials) with a lower energy than the original incident wave.

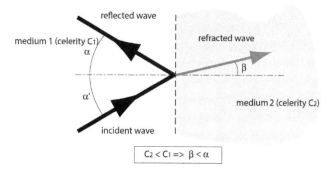

Figure 1.44. *Refraction of a sound wave (α = α')*

If the speed of propagation in the first medium is greater than the speed in the second medium, the angle of refraction is smaller than the angle of incidence or the angle of reflection.

1.6.8. *The Doppler effect*

This effect will be explained in section 5.4.2 of Chapter 5, which is dedicated to modulation effects, when we discuss the rotary effect.

1.6.9. *Beats*

The phenomenon known as a *beat* occurs when two sounds have similar frequencies. It is a direct consequence of the sensitivity of our auditory perception system. The alternating constructive and destructive interference of the sounds makes the volume of the sound signal seem to alternate between loud and quiet.

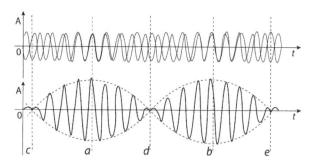

Figure 1.45. *The beat phenomenon. At a) and b), the signals are in phase, at c), d) and e), they are completely out of phase (antiphase)*

The frequency of the beat is equal to the absolute value of the difference of the two frequencies of each sound wave.

$$f_{beat} = |f_2 - f_1|$$

If the frequency of the beat is greater than 50 Hz, our brain is able to distinguish between the two sound sources. If not (<50 Hz), our brain detects a sound with an intermediate pitch (often called: the *subjective tone* or *combination tone*) with an intensity that appears to vary at the rate of the beat frequency.

1.7. Conclusion

Now that we have taken a brief tour around the notions of sound and acoustics, readers will have gained the foundations needed to understand the various sound effects that we shall discuss throughout the rest of the book.

We have not even nearly mentioned everything there is to know about sound. Interested readers can explore further using the web links and bibliography given at the end of this book, but you should now be able to continue reading without any difficulties, if you prefer.

2

Audio Playback

Multichannel sound, audio playback and audio media are important enough to have a dedicated chapter in this book. They often inspire many questions, many of which deserve to be answered.

The next few sections only give a simple description. They are not enough to fully understand the key ideas of multichannel processing, but offer a gentle introduction to the vocabulary, conventions and standards of audio.

Between 1964 and today, the world has gone through a variety of different media:

– the K7 audio cassette, invented by Philips in 1964;

– the laserdisc, or video disc, invented by David Paul Gregg in 1958, and introduced to the general public in 1972 by Philips and MCA (Music Corporation of America);

– the audio CD (compact disc), also invented by Philips in collaboration with the Japanese company Sony in 1979, based on laserdisc technology;

– the DVD (digital versatile disc), the result of collaboration between a large panel of companies specializing in audio (Philips, Toshiba, Time Warner, Matsushita Electric, Hitachi, Mitsubishi Electric, Pioneer, Thomson, etc.), launched in 1995;

– the Blu-ray disc (BD) and its competitor, the HD DVD[1], introduced in 2006, both specifically designed to support HD video.

---

[1] The HD DVD (high-density DVD) standard, originally created by Toshiba, was discontinued in March 2008.

The most recent of these media are also capable of storing multichannel music, notably including DVDs and Blu-rays, despite being primarily designed for video.

Even so, multichannel playback can still prove difficult. Multichannel systems require large setups with several speakers, which are often expensive, and ensuring that a given setup is compatible with the constantly expanding set of existing standards can be challenging. All in pursuit of perfect sound quality.

2.1. History

Before we begin to review the list of modern audio playback standards and techniques, it will be enlightening to start with a brief history of sound recording and reproduction.

In 1877, Edison[2] recorded the human voice for the first time on a sheet of tinfoil using a cylinder phonograph built by his best mechanic, John Kreusi[3]. He patented this invention on February 19, 1878, with patent number 200521.

Figure 2.1. *Edison's phonograph*

Thus, begins the history of sound recording. Chichester Bell and Charles Tainter, Charles Cros, Emile Berliner, Eldridge Johnson, Levi Montross, Gianni Bettini, Henri Lioret, François Dussaud and many others will help to perfect the invention.

---

2 Thomas Edison was born on February 11, 1847 in Milan, Ohio, and died on October 18, 1931 in West Orange, New Jersey. An American inventor and entrepreneur, he has 1,093 patents to his name.

3 John Kruesi, May 15 1843–February 22 1899, Swiss mechanic and engineer who worked closely with Edison to perfect a number of inventions.

The cylinder was superseded by flat discs made from shellac, then from celluloid a few years later. The first motorized phonographs were offered for sale in 1900.

In 1898, Poulsen[4] filed a patent in Denmark for a device that he named the "telegraphone". This device was the first magnetic recorder, and was based on the same principle as the phonograph, but used a steel wire instead of a cylinder, and a small electromagnet instead of a needle. The steel wire is the medium. After displaying his device at the *Exposition Universelle* in Paris in 1900, he was awarded American patent number 661619 for his invention.

Figure 2.2. *Poulsen's "telegraphone" based on a steel wire*

The first so-called "high-fidelity" (or hi-fi) discs were born in 1924, operating in the frequency band between 100 and 5,000 Hz.

As technology kept improving, the microphones used to pick up sound became more and more sensitive. In 1927, G. Neumann[5] allowed the whole industry to make a giant leap forward by inventing a completely new, highly sensitive microphone capable of perfectly reproducing the spectrum and range of recorded voices and instruments.

---

4 Valdemar Poulsen, November 23 1869–July 23 1942, Danish engineer and inventor.
5 Georg Neumann, 1898–1976, German inventor who founded the company Georg Neumann & Co. together with Erich Rickmann.

In the late 1920s, Fritz Pfleumer filed a patent for the use of magnetic oxide deposits on tapes made from paper or cellulose acetate. His company AEG[6] went on to build the first magnetic tape recorders. The Farben industrial group (BASF, Hoechst, Bayer) and Telefunken rapidly followed suit, creating the first magnetic tapes based on a plastic medium.

Figure 2.3. *AEG Magnetophon, 1935*

The speed of revolution of records was standardized for the first time in 1942, defined as 77.92 rpm in Europe and 78.26 rpm in the United States (commonly called 78-rpm records).

On June 21, 1948, vinyls were launched in the United States by teams led by René Snepvangers, and later Peter Goldmark, at CBS Laboratories[7]. This new medium was lighter, more robust and more faithful than 78-rpm records, despite having the same 12-inch diameter, and boasted a recording capacity of 30 min per side, a massive improvement on the 4.5 min of 78-rpm records. Vinyls were played at 33 rpm. Other sizes and speeds would be introduced later (7-, 10- and 12-inch 33 rpm; 7- and 12-inch 45 rpm created by RCA[8]; 12-inch 16⅔ rpm; and even 12-inch 8⅓ rpm and 4⅙ rpm).

---

6 Allgemeine Elektricitäts-Gesellschaft, German manufacturer of electronic and electrical equipment that also worked for the military and railway industries.
7 CBS Laboratories was founded in New York in 1936.
8 American company (Radio Corporation of America) founded in 1919 when General Electric decided to diversify into the radio sector.

By 1945, the quality of tape recorders had significantly improved, and they began to be more widely used.

However, we should bear in mind that sound playback systems were still monophonic at this point, although they had now advanced from a simple horn attached to a needle without any form of amplification in the first phonographs to an electronic audio system with amplifiers and loudspeakers. The latter were first developed by Siemens[9], who patented a moving-coil loudspeaker system on December 14, 1877. For a long time, his invention would be exclusively used for historical telephones, before being significantly improved and adapted for audio playback.

In the 1940s, the main concern was to achieve maximum gain, and people did not worry about distortion or bandwidth. Audio playback systems based on horns were supplanted by acoustic loudspeakers in the 1950s, but the latter still distorted the sound signal. The first speakers were large, cumbersome and often relatively ugly. Miniaturization would not be achieved until 1970s, when small speakers with moving elements and high displacement were introduced, finally capable of correctly reproducing low frequencies.

Stereophony was born in a series of experiments by Harvey Fletcher and most notably Alan Blumlein, who worked for EMI[10] in the United Kingdom. He realized that reproducing a wavefront was impossible, so he decided to try to build an audio playback system with two sound sources. He placed an observer at one of the vertices of an equilateral triangle, and sources at the other two vertices. During recording, he used the technique of *coincident microphones*. Stereo sound would require more than a decade to become popular, and the golden age of "high-fidelity" (hi-fi) would only begin in the early 1960s. However, movie theaters began using multiple channels and loudspeakers at the beginning of the 1950s. In 1959, radio took its first steps toward stereo broadcasting, but stereo television would not see the light of day until 1978, in Japan.

Stereo and hi-fi reached their peak at the turn of the 1970s, and the industry began to search for new ideas, ultimately leading to *quadriphonic sound*. Recording was simple enough, but reproducing the sound was another story: how can you fit four channels onto a vinyl that is already struggling to store just two? A number of solutions based on complex matrix decoding systems were attempted, but the results

---

9 Ernst Werner von Siemens, December 13 1816–December 06 1892, German inventor and industrialist, founder of the company bearing his name.

10 EMI (Electrical and Musical Industries), a major record company, was founded in 1931 by merging Columbia Records and HMV (His Master's Voice).

were ultimately disappointing, and the general public remained unimpressed by the expensive and highly fragile equipment.

In 1971, Stanley Kubrick used a new noise reduction technique named "Dolby[11] A" for the first time in his film "A Clockwork Orange".

In 1978, Dolby Laboratories announced its Dolby Stereo system. This system, specially designed for movies, condensed four separate audio channels (left, center, right and surround) onto just two channels using a matrix decoding system. It also incorporated a noise reduction system that improves the dynamics of the sound. For the first time, viewers could feel immersed in space while listening, even though this system seems archaic to us today. In 1982, it was made available to the general public in the form of *Dolby Surround*, followed by an improved version, Dolby Surround Pro-Logic (DPL), in 1987. From this point onward, multichannel systems flourished, especially in movie theaters and private homes, with the rise in popularity of home theaters. Digital audio proved immensely popular, with *Dolby Digital 5.1* in 1992, the *Digital Theater System* (DTS) in 1992, 6.1-channel standards in 1999 and so on. Today, a niche market has developed for *DVD-Audio* and *Super Audio CD* (SACD), specifically dedicated to music recordings, although audio CDs continue to dominate.

2.2. Dolby playback standards and specifications

The Dolby Stereo, Dolby Surround and DPL formats refer to specific decoding techniques. The signal is encoded as Dolby Surround in each case, also known as *Dolby MP Matrix* (MP for Motion Picture).

2.2.1. *Dolby Surround encoding and decoding*

The objective of this type of encoding is to deliver four sound channels, two for the front left and right, one for the front center and one for the rear.

The center channel is attenuated by 3 dB, then added to the left and right channels. If the signal is only in the center, the left and right channels are identical.

---

11 Named after its inventor, Ray Dolby, born in Portland, Oregon, an American engineer and physicist who invented a noise reduction system in the 1960s to improve the audio quality of magnetic recordings.

The rear channel is also attenuated by 3 dB, then the signal is filtered to only include frequencies between 100 Hz and 7 kHz. After filtering, the Dolby B noise reduction system is applied to a specific frequency range.

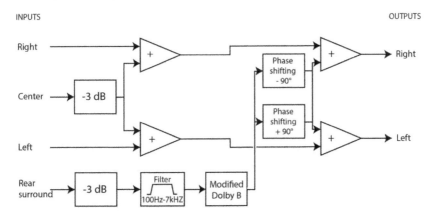

Figure 2.4. *Flowchart of a Dolby Surround encoder*

Decoding is executed passively. The decoder applies the inverse operation to the encoder. However, a delay is added when decoding the rear channel, typically between 10 and 30 ms. Because of the Haas effect (see section 1.2.3), this reduces the listener's awareness of the residual left and right components that still remain in the surround channel. A 7-kHz low-pass filter is also applied to this channel to only retain high frequencies, which are easier to identify for our ears.

2.2.2. *Dolby Stereo*

Figure 2.5. *The Dolby Stereo logo*

Dolby Stereo was developed in the 1970s and is primarily designed for the movie industry. It delivers four audio channels. As well as the two front channels, which recreate a stereo signal, there is an additional center channel and a rear channel. This rear channel is typically reserved for ambient sounds and sound effects.

2.2.3. Dolby Surround

Figure 2.6. *The Dolby Surround logo*

Dolby Surround (home theater version) and *Dolby SR* (*Spectral Recording* – movie theater version) provides four channels with a greater depth of field than Dolby Stereo: two front left and right channels (stereo), a third center channel and a fourth rear channel. Music and sound effects are played through the stereo front channels, which are placed on either side of the screen. The center channel is reserved for dialogue. It is typically placed behind the screen. The rear channel, filtered to the frequency band 100 Hz–7 kHz, reinforces the sensation of spatial depth by playing ambient sounds and sound effects. This analog system, released in 1982, is still used today in some movie theaters. Even if the playback system does not have a Dolby SR decoder, the stereo signal can simply be played directly. This format is described as 4.0, i.e. four sound channels without a dedicated channel for low frequencies.

Figure 2.7. *The Dolby Surround playback symbol*

2.2.4. Dolby Surround Pro-Logic

Figure 2.8. *The Dolby Surround Pro-Logic logo*

DPL, introduced in 1987, is a special method for decoding Dolby Surround. All monophonic signals are sent to the center speaker, and all signals in antiphase are sent to the rear channel. The center and rear signals are subtracted from the front left and right channels. A 120-Hz filter is applied to redirect low frequencies to a dedicated speaker.

2.2.5. Dolby DIGITAL AC-3

Figure 2.9. *The Dolby Digital logo*

This format was very widely used until around 2010. It was developed in 1991, and was made available to the general public in 1995 on NSTC laserdiscs. It was the first fully digital multichannel sound format. It could be found in many movie theaters and was also widely used by DVDs.

The Dolby Digital encoding scheme is known as AC-3 (Audio Coding 3). This process only encodes frequencies that are audible to the human ear, which minimizes the volume of sound data that needs to be stored. Dolby Digital supports three different sampling rates, 32, 44.1 and 48 kHz, with transfer rates ranging from 32 to 640 kbps (kilobits per second).

This type of encoding is called 5.1, with five sound channels plus one extra channel dedicated to low frequencies. All of these channels are independent of the others:

– two front channels for music and sound effects (20 Hz–20 kHz);

– one front center channel for dialogue (20 Hz–20 kHz);

– two rear left and right channels for ambient sounds and effects (20 Hz–20 kHz);

– one so-called low-frequency effects (LFE) channel to reinforce low sounds (typically between 3 and 120 Hz). This channel is usually played through a special speaker known as a "subwoofer".

All six channels are independent, unlike previous Dolby systems encoded on two tracks.

Dolby Digital also supports mono, for example when storing old film that has been converted.

Dolby SR-D is the digital version of Dolby SR, reserved for movie theaters, and is equivalent to Dolby AC3 for home theaters.

Figure 2.10. *The Dolby Digital playback symbol*

2.2.6. Dolby Surround EX

Figure 2.11. *The Dolby Digital logo*

The EX (Extended) format was launched in 1999, and extends the specifications of Dolby Digital by adding an additional rear center channel. It was developed by a collaboration between Dolby Laboratories and Lucasfilm THX.

Figure 2.12. *The Dolby Digital EX playback symbol*

2.2.7. Dolby Surround Pro-Logic II

Figure 2.13. *The three logos of Dolby Surround Pro-Logic II*

Launched in 2000, this format replaces DPL, achieving better sound quality than its predecessor by decoding five channels from a stereoencoded signal: front left, front right, center, rear left and rear right. Another advantage is that it can process any stereo signal, whether encoded in Dolby Surround or not. This allows a multichannel signal to be derived from any audio source, allowing end users to make the most of their home theater. The matrix decoding process of the original signal was also improved, providing excellent separation between channels, and increasing the bandwidth to 20 Hz–20 kHz on the rear channels.

This format has two modes: Movie mode and Music mode, adapted to each type of media.

In Movie mode, the source needs to have been encoded in Dolby Surround. The rear surround channels are stereo, and the bandwidth is said to be "full", meaning that it is not limited to 7 kHz like DPL.

The Music mode can process any stereo music source, whether encoded in Dolby Surround or not. This mode can be configured with three criteria:

– the *center width*, i.e. the balance of left and right that is sent to the center channel;

– the *dimension*, which adjusts the front-to-rear balance;

– the *panorama*, which creates identical sound on the front and rear channels.

This encoding could be used with VHS sources, although it was primarily designed for movies.

REMARKS.– There are two other version of Dolby Surround Pro-Logic II, *x* and *z*.

Pro-Logic IIx retains the same characteristics, with the additional option of playing 6.1 or 7.1 from a stereo or 5.1 source. It has three playback modes, "movie", "music" and "game", to optimize sounds effects depending on their nature.

Pro-Logic IIz adds an extra spatial dimension, allowing two surround speakers to be placed above the front speakers. This introduces a new vertical component that was absent from other encodings when this format was launched (for example sound can now come from the roof, the sky, the attic, etc.).

Figure 2.14. *The Dolby Pro Logic IIx and IIz playback symbols*

2.2.8. Dolby Digital Plus

This encoding was designed for consumer HD applications. The quality was improved, and the number of available channels was increased. This is the standard used by all HD DVDs, and is often used for Blu-rays. It works in 7.1 and delivers sufficiently high transmission rates to support television and streaming at 6 Mbps (megabits per second).

Figure 2.15. *The Dolby Digital Plus playback symbol*

2.2.9. Dolby TrueHD

It is one of the latest forms of multichannel encoding, developed for HD. This standard uses lossless technology, with a bitrate of 18 Mbps. Dolby TrueHD supports 14 channels at a resolution of 24 bits and 96 kHz.

Many Blu-ray films are encoded in this format.

Figure 2.16. *The Dolby TrueHD playback symbol*

Audio Playback 55

2.2.10. *Dolby Atmos*

Like Dolby Pro-Logic IIz, this format adds a vertical spatial component to the sound, with the difference that it requires multiple elevated speakers (e.g. on the ceiling). Dolby Atmos supports up to 64 channels.

This format is often used in movie theaters. Each surround speaker can play different parts of the soundtrack, provided that the original mix was configured accordingly.

Figure 2.17. *One of the Dolby Atmos playback symbols*

Dolby Atmos can be used in compatible home theaters (the amplifier must recognize the standard and the speakers need to be correctly placed). There are several possible speaker configurations.

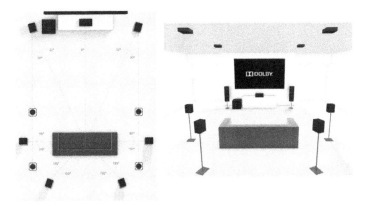

Figure 2.18. *One example home theater speak configuration for Dolby Atmos 7.1.4 (12 speakers) (source: www.son-video.com)*

2.3. DTS encodings

DTS is an encoding that competes with Dolby Digital. It has also evolved over time, and multiple different standards have emerged.

2.3.1. DTS

Figure 2.19. *The DTS logo*

DTS was released in 1999, and encodes six channels, like Dolby Digital (5 + 1 LFE). However, DTS has a lower compression rate than Dolby Digital and therefore offers better quality. In movie theaters, this system uses time codes to synchronize the audio, which is recorded on a separate CD-ROM, and the image on the film. Many movie theaters around the world still support this format, which has been available since 1993.

DTS supports multiple sampling rates ranging from 8 to 192 kHz. The transfer rate varies from 32 to 4,096 kbps.

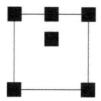

Figure 2.20. *The DTS playback symbol*

2.3.2. DTS Neo 6

Figure 2.21. *The DTS Neo 6 logo*

Released in July 2000, this format has two modes, Movie and Music. It was developed by DTS as a direct competitor for DPL II.

2.3.3. DTS ES 6.1

Figure 2.22. *The DTS ES logo*

This configuration of DTS has an extra channel, designed to be placed at the center rear. It is a director competitor of Dolby Digital Surround EX.

2.3.4. DTS 96/24

The goal of this standard is to improve the quality of the signal by using a 96-kHz, 24-bit encoding on six channels, with the option of outputting the digital stream via S/PDIF[12].

Figure 2.23. *The DTS 96/24 logo*

2.3.5. DTS HD Master Audio

This is a lossless format that uses a variable bitrate technique to produce high-quality signals. It supports two encoding frequencies, 96 kHz for eight audio channels (7.1), and 192 kHz for six channels (5.1).

The bitrate of this type of DTS ranges from 1 to 24.5 Mbps.

Figure 2.24. *DTS HD Master Audio logo*

---

12 Sony/Philips Digital Interface (S/PDIF) is a format for digital audio data transmission, defined in 1989.

REMARK.– There is an intermediate standard called DTS HD High-Resolution Audio that delivers 7.1 at 96 kHz, with bitrates ranging from 2 to 6 Mbps.

Figure 2.25. *The DTS High-Resolution Audio logo*

2.3.6. *DTS X*

DTS X supports up to 32 speakers, but it is usually used with 11.2-channel systems. It has a sampling rate of 96 kHz.

The goal of this standard is to faithfully recreate a 3D audio space by introducing the notion of independent sound objects that can move and occupy positions throughout this space.

Figure 2.26. *The DTS X logo*

DTS X is built around DTS HD Master Audio. In the most general case, this standard adds two or four additional speakers to 5.1, 7.1 or 9.1 configurations. These extra speakers are placed above the front speakers and rear surround speakers.

Of course, like all of the more modern standards, DTS X requires a compatible playback system.

2.4. Special encodings

As well as the Dolby and DTS standards listed above, both have other, more specific encodings:

– Dolby Headphone: introduced in 2000, this standard allows surround environments to be recreated within a conventional stereo headset;

– Dolby Virtual Speaker: introduced in 2000, this standard creates the illusion of surround sound using a single pair of frontal speakers;

– DTS Headphone: X: launched in 2013 to compete with Dolby Headphone, this standard simulates surround sound with headsets;

– DTS Interactive: This a version of DTS designed for video games.

REMARK.– Even the long list of Dolby and DTS standards presented above is not exhaustive. There are other variants designed for specific environments and hardware, and even other experimental standards.

2.5. SDDS

Figure 2.27. *Logo of the SDDS format*

Dating back to 1994, the SDDS (Sony Dynamic Digital Sound) format is exclusively reserved for movie theaters. It was developed by Sony and Columbia. It is usually arranged in 5.1 or 7.1 configurations. It uses a proprietary data compression algorithm developed by Sony for the Sony MiniDisc, called ATRAC (Adaptative TRansform Acoustic Coding).

2.6. THX certification

Figure 2.28. *The logo of THX (Tom Holman eXperiment)*

The acronym THX comes from the name of one of the engineers at Lucasfilm, Tomlinson Holman, who wrote specifications for reproducing movie soundtracks according to a set of very strict quality criteria in response to a request by George Lucas. The name is also a reference to George Lucas's first film, "THX 1138", which was released in 1971. The first film shown in theaters under this standard was "The Return of the Jedi" from the Star Wars saga.

THX recommendations are now also available in a form suitable for home theater.

The main areas considered by THX quality certification are:

– sound balance;

– sound power;

– high dynamic range;

– low background noise;

– realistic sound space;

– immersive surround sound;

– realistic distribution of sound perspectives;

– clarity and intelligibility of dialogue.

The surround speakers of THX systems are bidirectional (dipole speakers), and the frequency band between 80 and 20 kHz is covered by five speakers, reserving low frequencies for subwoofers (from 1 to 3 for THX Home Cinema systems) on the LFE channel.

The standard also includes other specifications relating to the theater venues themselves, such as room reverb, acoustic insulation, viewing angles for spectators and projection type.

The hardware requirements are very strict for both commercial and home theaters, placing requirements on the media (LD, DVD, etc.), as well as the connection cables and materials used by the sound system.

REMARK.– There is also a separate THX I/S Plus standard designed for small home theaters.

2.6.1. *THX select and ultracertification*

Figure 2.29. *The logos of THX Select and Ultra*

THX Select applies to lower range sound systems, namely systems with power below 100 W per channel, which is a minimum requirement of regular THX. To differentiate between the two more clearly, Lucasfilm rebranded THX-certified systems as THX Ultra when THX Select was launched.

2.6.2. *THX Ultra 2 certification*

Figure 2.30. *The logo of THX Ultra 2*

THX has of course evolved over time to keep up with the introduction of new standards. In 1995, THX 4.0 was replaced by THX 5.1. With the arrival of 6.1 standards, THX Surround EX and THX Select Surround EX were born. THX Select/Ultra 7.1 was developed after the introduction of 7.1.

However, THX Ultra 2 is a particularly noteworthy step forward, introducing support for DVD-Audio and SACD, which are specially designed for listening to music.

2.7. Multichannel audio recording

To properly understand this type of audio recording, we need to revisit some of the principles of stereophonic audio. The challenge is to record a homogeneous audio signal by accounting for the acoustic properties of the recording environment.

We will not go into extensive detail here – we are moving away from the central theme of this book, and I do not specialize in this type of recording. If you are interested, you can explore the internet links and bibliography at the end of the book to find out more.

However, a few things are worth mentioning.

Sound recording techniques can be based on three, four or five microphones, or sometimes even more.

One key principle is that there is one essential recording axis along which the primary audio image of the scene is concentrated. This axis defines the global spatial configuration of the sources by perfectly balancing the direct sound field and the secondary sound field reverberated by the environment. It establishes the foundation for the timbre and determines the rendering of the recording by providing the spatial localization of each source.

In addition to this primary audio image, there is a secondary, less concentrated image containing ambient sounds without any direct sources. This image represents the atmosphere and soundscape of the audio scene, and is usually constructed from the primary image.

The primary and secondary images are supplemented by the front audio, as well as any very close sounds surrounding the listener.

Multichannel recording is based on two main factors, localization and spatialization. The goal is to combine both of these factors to cover a 360° space without any gaps, while keeping *cross-talk* at acceptably low levels.

Many recording setups meet these requirements: the Decca tree, the MMAD (Multiphonic Microphone Array Design), Double ORTF, the Hamasaki square, OCT (Optimized Cardioid Triangle) Surround, Soundfield, the IRT cross, the holophone H2, etc.

Figure 2.31. *An IRT cross recording setup*

These systems are of varying effectiveness and produce very different sound renderings. Thus, there is no single universal solution to the problem of multichannel recording, but rather multiple solutions depending on the location and the desired levels of ambient sound, as well as the desired sensitivity and approach chosen by the sound technician.

2.8. Postproduction and encoding

Postproduction mixing, encoding, decoding, measuring and control of multichannel formats requires highly specialized tools, typically integrated into a DAW. A wide variety of tools are also available as racks developed for studios that specialize in this type of mixing/encoding/mastering.

Table 2.1 provides a simple list of plugins for manipulating and processing multichannel audio. Many of these plugins were designed to be integrated into the Avid Protools software. The others have various other formats (VST, VST2, VST3, AU, etc.) and can be used with various DAWs and digital audio editing suites. More information about the specifications of each of these tools is easy to find online.

If you wish to find out more about audio processing systems like Dolby, DTS, etc., a few useful references are listed in the web links and bibliography sections of this book.

Editor	Name	Remarks
Avid	Dolby Surround Tools	Surround encoding and decoding
Waves	360° Surround Tools	Toolbox for 5.1 surround production: compressor, quantizer, limiter, low-pass filter, reverb, imager, mixer, VU meters, etc.
Waves	DTS Neural Surround Collection	Set of three tools for encoding 5.1/7.1 DST
Auro Technologies	Auro-3D Authoring Tools	Surround encoder
New Audio Technology	Spatial Audio Designer	5.1/7.1 surround encoder
Dolby	Atmos Mastering Suite	Dolby Atmos surround encoder
Dolby	Atmos Production Suite	Toolbox for Dolby Atmos surround production
Dolby	Media Emulator	Metadata simulator for 5.1 Dolby playback – Dolby E preencoder
Dolby	Surround tools	Surround encoder/decoder
Cycling'74	UpMix	Toolbox for 5.1 surround encoding
Neyrinckl	SoundCode for DTS	Surround encoder/decoder
DTS	Neural Upmix	UpMix – stereo to 5.1 – 5.1–7.1
Sonnox	Fraunhofer Pro codec	5.1 encoder
Sonic Anomaly	Surround Pan	Freeware – DTS encoder
Minnetonka audio	SurCode DTS	DTS encoder/decoder
Minnetonka audio	SurCode Dolby	Dolby encoder/decoder
Dolby	LM100	Software version of a loudness measuring tool
Neyrinck	Mix 51	5.1 mixing tool
Neyrinck	V-Mon	10 inputs to 5.1 measuring tool – 7.1 output control
Neyrinck	Soundcode LtRt	LtRt Dolby Pro-Logic I and II encoder
TC Electronic	LM5D	Tool for measuring and loudness control

Table 2.1. *Examples of tools for mixing, encoding, decoding, controlling and measuring multichannel 5.1/7.1 Dolby or DTS audio*

2.9. Multichannel music media: DVD-Audio and SACD

DVD-Audio and SACD were specifically designed to support 5.1 multichannel music playback. Although the idea might seem extremely promising, these media struggled to conquer the market and sales have stagnated. However, these two formats are, nonetheless, worth bearing in mind. They can be used to take advantage of audio with spatial depth in home theater systems whenever a compatible media player is available.

2.9.1. *DVD-Audio*

In 1996, manufacturers and a group of experts from the DVD Forum began to draft specifications and develop a new DVD-based musical media format. In May 1998, the DVD-Audio standard (usually referred to as DVD-A) was unveiled. It was finalized in February 1999. This standard introduces a new lossless compression process for audio data called MLP (Meridian Lossless Packing) that can achieve up to 45% reduction in space and which supports real-time decompression.

Figure 2.32. *The DVD-Audio logo*

DVD-Audios have the same physical dimensions as audio CDs and use the same encoding technique, namely Pulse Code Modulation (PCM).

The main difference lies in the sampling rate. CDs have a fixed sampling frequency of 44.1 kHz at 16 bits, whereas DVD-A offers the choice between 44.1, 88.2, 96, 176.4 and 192 kHz at 16–24 bits.

Table 2.2 shows the supported combinations of channels and sampling frequencies.

The bandwidth can thus range up to 96 kHz (192 kHz for 24 bits), which is much higher than the 20 kHz offered by CDs. The signal-to-noise ratio is 144 dB.

Number of channels	Sampling frequency at 16, 20 or 24 bits					
	44.1 kHz	48 kHz	88.2 kHz	96 kHz	176.4 kHz	192 kHz
1.0 (mono)	OK	OK	OK	OK	OK	OK
2.0 (stereo)	OK	OK	OK	OK	OK	OK
2.1 (stereo)	OK	OK	OK	OK	–	–
3.0/3.1 (stereo + mono surround)	OK	OK	OK	OK	–	–
3.0/3.1 (3-stereo)	OK	OK	OK	OK	–	–
4.0/4.1 (3-stereo + mono surround)	OK	OK	OK	OK		
5.0/5.1 (surround)	OK	OK	OK	OK		

Table 2.2. *Supported sampling frequencies and channels*

Depending on the sampling frequency, the number of channels can be modulated. For example, 5.1 playback can be achieved by three 96-kHz 24-bit channels and two 48-kHz 20-bit channels (see Table 2.2).

DVD-Audios can also store other types of information than just soundtracks, such as text or images. Each DVD-A can be divided into nine groups of 99 tracks, each of which can have 99 indexes. Depending on the type of DVD, audio recordings can last up to 7 h of 16-bit 44.1-kHz stereo, or 74 min of six 24-bit 96-kHz channels (DVD-5[13]). Type-9 DVDs and above have even higher storage capacity. Mixed audio and video DVDs are also possible.

DVD-Audios require a compatible DVD player.

---

13 There are several types of DVDs: DVD 5 – 4.7 GB – single-sided/single-layer, DVD-9 – 8.54 GB – single-sided/double-layer, DVD-10 – 9.4 GB – double-sided/single-layer, DVD-14 – 13.24 GB – (DVD-5 + DVD-9) side 1: single-layer/side 2: double-layer, DVD-18 – 17 GB – double-sided/double-layer.

2.9.2. Super Audio CD

In 1998, Philips and Sony announced a new medium with the same physical dimensions as CDs, termed as SACD. This new medium supports high-quality music recordings with multichannel spatial sound.

Figure 2.33. *The SACD logo*

This type of media can be played by dedicated SACD players (which are also compatible with CDs), SACD-compatible DVD players and Direct Stream Digital (DSD) players.

The technology on which SACDs are based does not use the same PCM process as DVD-Audios and CDs. The audio sampling frequency is 2,822.4 kHz, which is 64 times the sampling frequency of audio CDs (44.1 kHz).

However, audio CDs are encoded with 16-bit samples, whereas SACDs have a single bit, which explains the extremely high frequency. The 16 bits of audio CDs are used to define 65,536 different values that characterize frequencies between 20 Hz and 20 kHz. With only 1 bit, the only possible values are 0 and 1, and so a different method is required. The value 1 is interpreted as a frequency increase, and the value 0 is interpreted as a frequency decrease. This is repeated 2,822,000 times per second (sampling frequency of SACDs). This technique is advantageous because consecutively sampled frequencies have very similar values, resulting in a loss in the sampled signal that is extremely low relative to the original signal. This technique is called DSD.

SACDs have much higher bandwidth than audio CDs, ranging from 1 Hz to 100 kHz, i.e. much wider than the spectrum perceived by the human ear. However, we should bear in mind that some frequencies can influence the sound even though they are inaudible. For example, infrasounds can be physically felt by the body. The signal-to-noise ratio of SACDs can range up to 120 dB.

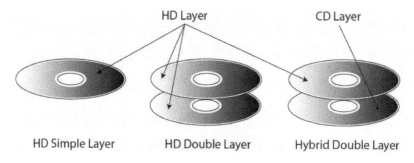

Figure 2.34. *Three types of SACDs*

There are three types of SACDs:

– single-layer high density (HD), which can only be read by SACD players;

– double-layer HD, which can also only be read by SACD players;

– hybrid double layer with two different layers. One of the tracks is reserved for SACD players, and the other is reserved for conventional CD players.

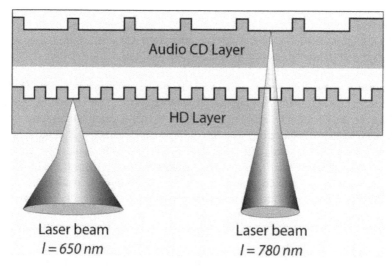

Figure 2.35. *A hybrid double-layer SACD. The laser beam on the right, with a wavelength of 780 nm, passes through the HD layer, which is transparent to it. The laser beam on the left, with a wavelength of 650 nm, can read the HD layer*

2.9.3. Comparison of CDs, SACDs and DVD-Audios

Table 2.3 summarizes the most important features of these three media.

Type	CD	SACD	DVD-A
Diameter (cm)	12	12	12
Sampling rate	44.1 kHz 16 bits	2,822.4 kHz 1 bit	44.1–192 kHz 16–24 bits
Bandwidth	20 Hz–20 kHz	100 kHz	96 kHz
Signal-to-noise ratio	96 dB	120 dB	144 dB
Channels	2 (stereo)	5.1	5.1
Notes	–	One audio CD track for the hybrid double-layer version	Can store text or images
Storage capacity	700 MB	4.7 GB	4.7 GB

Table 2.3. *Comparison of CDs, SACDs and DVD-As*

REMARK.– There are also next-generation versions of these standards, most notably of the SACD, with the SHM SACD (Super High Material Compact Disc) created in 2010, and the DSD–CD.

Figure 2.36. *Logos of the SHM SACD and the DSD–CD*

2.10. Conclusion

As we can see, there are a large number of existing standards for multichannel encoding. New additions are constantly being made over time as media and playback technology changes. It is easy to get lost, even before considering the options for certifying and configuring audio spaces (large and small theater venues, home theater setups, etc.).

The DVD-A and SACD standards are still current, although the market for these media remains very limited. Today, they are almost exclusively used by audio enthusiasts who are passionate about classical music.

Properly encoding audio and complying with the subtleties of each standard can represent a significant challenge for recording studios. Acquiring the proper equipment for encoding and playback can quickly present substantial technical and financial challenges.

3

Types of Effect

Before we begin the central descriptive part of this book on sound effects, we will take a moment to describe and classify the various types of effect that we will study in more detail later.

We will consider two criteria: physical appearance and audio processing.

We can divide effects into three categories according to their physical appearance: rack equipment, pedals and plugins (software components).

From the perspective of audio processing, we can define eight broad families:

– filtering effects;

– modulation effects;

– frequency effects;

– dynamic effects;

– time effects;

– unclassifiables.

In the following sections, we will review the various aspects of each criterion.

3.1. Physical appearance

First, we will consider the physical hardware and software forms in which effects can exist.

3.1.1. Racks

Rack effects are primarily used for sound recording in studio environments, mixing, mastering and sound reinforcement, typically in connection with a mixing desk.

Audio racks are analog or digital effects integrated into compact box-shaped units that can be vertically stacked inside a cabinet with standardized shelves designed specially to accommodate them.

Strictly speaking, the rack refers to the cabinet itself, which stores "rack-mountable" effects units. However, over time, the effects themselves have come to be known as racks.

3.1.1.1. Standards and dimensions of racks

Racks (storage shelves in cabinets) can be mounted on wheels in a road case ready for live applications, or permanently installed in a recording studio. Their dimensions are regulated by the standards IEC 60297[1] and EIA 310-D[2].

By default, professional audio racks have a width of 19 inches (482.60 mm). They have two vertical metal rails, each 0.625 inches (15.875 mm) wide, with 17.75 inches (450.85 mm) of space in between.

Figure 3.1. *Example of a 19-inch audio rack cabinet*

Holes are drilled in the rails, with 18 and 5/16 inches (465.14 mm) of horizontal space and 0.625 inches (15.875 mm) of vertical space between holes.

---

1 International Electrotechnical Commission. This standard was formerly DIN 41 494 (see web links at the end of the book).
2 Electronic Industries Association.

The holes are often square-shaped, designed for captive nuts (also known as caged nuts). Some rails have prethreaded holes. Two types of thread are used: 10–32 and 12–24. The nut diameter is typically 6 mm.

Figure 3.2. *Captive nuts (or caged nuts) and screws*

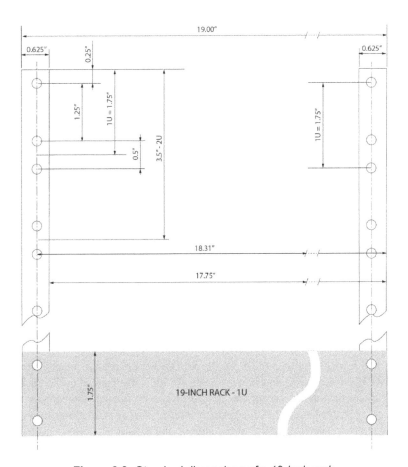

Figure 3.3. *Standard dimensions of a 19-inch rack*

The height of rack-mountable modules is measured in units of U, which is simply short for unit. One U is equal to 1.75 inches (44.45 mm).

The height of a module is typically given in multiples of U, e.g. 1U, 2U, 3U, 4U or more (Eurocard system).

To ensure that modules can be easily mounted, a clearance of 1/32nd of an inch (0.79 mm) is left between modules, resulting in a face height of 43.66 mm for each module.

3.1.1.2. Effects units

Let us now move on to the actual contents of the rack, which for us means the musical audio effects themselves. Effects units are usually reasonably compact, rarely exceeding 3U in size.

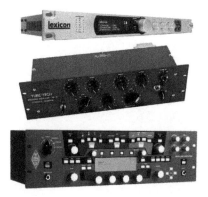

Figure 3.4. *Three examples of (19") rack-mountable effects with different heights: "PCM92" reverb by Lexicon (1U), "PE-1C" parametric equalizer by Tube-Tech (2U) and the "Profiler" multieffects processor by Kemper Amps (3U)*

The audio inputs/outputs (Jack, XLR, S/PDIF, optical, etc.), other connections (USB, MIDI, RS422, etc.) and power cables are typically located at the back of the unit. Extra connectors are sometimes placed on the front, such as headphone jacks, remote control interfaces and microphone inputs. These front connectors can, for example, be redundant duplicates of the ones on the back.

3.1.2. Pedals

Pedal effects take the form of a physical pedal, sometimes in combination with other equipment like a rack.

3.1.2.1. *Amplifier pedals*

Some effects are integrated directly into an amplifier and are operated using a convenient pedal that connects to a dedicated input on the amp.

Figure 3.5. *Fender amp, 65 Deluxe Reverb model (top left), with its effects pedal (top right) for controlling the vibrato and reverb effects. The jack socket for connecting the pedal is visible in the bottom image ("foot switch")*

Usually, with the occasional exception, these pedals have limited functionality, and are simply used to toggle tremolo, vibrato or reverb effects that must be preconfigured using the controls directly on the amp.

3.1.2.2. *Dedicated effects pedals*

Since the first electronic guitar effects began to appear in the mid-1970s, engineers and sound technicians have invented control designs that can be easily operated while standing or sitting. Pedals are a perfect example of this.

REMARK.– Effects pedals were inspired by the earlier pedals used to control the volume of electronic organs.

Before the introduction of these pedals, effects such as distortion or *overdrive* needed to be configured via the controls on the amplifier, which was highly impractical on stage.

Effects pedals are connected between the instrument and the amplifier, and can be chained together to allow multiple pedals to be combined or used independently.

Some musicians also use *pedalboards*, which make it easier to manage audio connections and power supplies with multiple pedals.

Figure 3.6. *Two pedalboards: the "Pro series Pedalboards" by Trailer Trash and the "BCB-60" by Boss*

The controls for configuring or activating the effect can, for example, be manual switches and buttons, or alternatively a mechanical (potentiometer + rack-and-pinion) or optical lever system that enables the parameters of the effect to be continuously varied by tilting the pedal.

REMARK.– Pedals based on tilting lever systems often include a switch or other mechanism located under the moving part of the pedal to toggle the effect on or off.

Figure 3.7. *Effects pedals with lever-and-switch systems. A wah-wah pedal "V847" by Vox (top). The rack-and-pinion, potentiometer and the actuator switch are visible under the footrest. An echo-delay pedal "Memory Man with Hazarai" by Electro Harmonix (bottom). It has two switches and various knobs for settings.*

REMARK.– Although pedals are usually designed for guitar and bass, they can also be used with many other instruments, such as organs, electronic keyboards (e.g. Fender, Wurlitzer, etc.), violins and so on.

3.1.2.3. Racks with pedals

Racks are often more powerful and flexible for musicians. In some cases, they can be placed on top of the amplifier and controlled via foot pedal. Most are capable of saving presets in their memory. These presets can then, for example, be selected by scrolling through them.

Figure 3.8. *Example of a multieffects rack designed for guitar, with a control pedal. This model is the G-sharp by TC Electronic*

Many effects racks behave like emulators, recreating specific amplifiers or certain popular effects that can be found on the market. These racks are often suitable for studio usage.

Figure 3.9. *A rack for emulating amplifiers and effects, the "Eleven" model by Avid. It can be used either in the studio or live directly with guitars or other instruments. Below it is a control pedal compatible with the MIDI standard, the "Ground Control Pro" by Voodoo Lab.*

3.1.3. Software plugins

Plugins are software components recognized by a Digital Audio Workstation (DAW) such as Avid Pro Tools, Magix Samplitude Pro X, Cakewalk Sonar, Presonus Studio One, Apple Logic Pro, Ableton Live, and many more, or alternatively by digital audio processing software such as Adobe Audition, Magix Sound Forge, Steinberg Wavelab, Audacity and so on. This list is far from exhaustive.

The format of plugins varies depending on the operating system and the software.

Table 3.1 lists the characteristics of each plugin format. Users can see the format of a plugin by looking at its file extension.

File extension	Editor	Software–hardware compatibility
VST (Virtual Studio Technology)	Steinberg 1996	DAW–Audio processing software–PC/Mac
VST2	Steinberg	DAW–Audio processing software–Improved version of VST–PC/Mac
VST3	Steinberg	DAW–Audio processing software–Improved version of VST2–PC/Mac
AU (AudioUnits)	Apple	DAW–Audio processing software–Mac–OSX and Mac OS
AAX (Avid Audio eXtension)	Avid	Pro Tools 64 bits–PC/Mac
MAS (MOTU Audio System)	MOTU (Mark of The Unicorn)	Digital Performer–Mac
AS (Audiosuite)	Digidesign	Pro Tools–PC/Mac
EASI (Enhanced Audio Streaming Interface)	Apple	Logic Audio–Mac
RTAS (Real-Time Audio Suite)	Digidesign	Pro Tools 10 minimum–PC/Mac
TDM (Time Division Multiplexing)	Digidesign	Cartes HD Core or HD Accel Digidesign–PC/Mac
DX (Direct X)	Microsoft	PC

Table 3.1. *Plugins by file extension*

Many plugins exist in multiple versions to provide compatibility with multiple types of software or hardware.

3.2. Audio processing

As we mentioned at the beginning of the chapter, there are eight families of audio processing effects that we can use as a classification. Chapters 4–9 describe the effects that belong to each of these families.

Table 3.2 gives a list of effects by family.

Family	Effect
Filtering	Graphic equalization
	Parametric equalization
	Semiparametric equalization
	Band-stop equalization
	Linear phase equalization
	Dynamic equalization
	Wah-wah
	Auto-wah
	Crossover
Modulation	Flanger
	Phaser
	Chorus
	Rotary, univibe, rotovibe
	Ring modulation
Frequency	Vibrato
	Transposer
	Octave up/down
	Pitch shifter
	Harmonizer
	Autotune
Dynamic	Compressor
	Limiter
	Expander
	Noise gate
	De-esser
	Saturation
	Fuzz
	Overdrive
	Distortion
	Exciter, enhancer, embellisher
Time	Analog reverb
	Digital reverb
	Convolution reverb
	Delay
	Slapback
	Doubler
	Pan delay
	Echo
Unclassifiables	Fuzzwha
	Octafuzz
	Shimmer
	Tremolo
	Declicker
	Decrackler
	Denoiser
	Declipper
	Debuzzer
	Looper
	Time stretching
	Resampling
	Spatialization

Table 3.2. *Families of effects*

3.3. Conclusion

This brief introduction to the various existing types of effect will make the following chapters easier to understand. We will examine most of the sound effects that we might find in a studio, on stage and on live sets.

The order in which we will study these effects has a certain logic that will allow us to discuss the connections and combinations between effects, but any readers who are interested in a specific effect are welcome to skip directly to the relevant section. References and hints are given wherever they are helpful.

4

Filtering Effects

Filters can be used to cut, eliminate, boost or modify regions or frequency bands of arbitrary width in an audio signal. Some filters are extremely selective, only targeting very narrow frequency components, e.g. based on a gain threshold.

4.1. Families of filtering effects

There are four large families of filtering effects:

– *high-pass filters*: these filters only allow through frequencies above a certain frequency threshold, called the cut-off frequency. They are used to reduce low frequencies;

– *low-pass filters*: these filters are the reverse of high-pass filters. They only keep frequencies below the cut-off frequency;

– *band-pass filters*: only the frequency band between an upper threshold and a lower threshold is allowed through. These filters are often used to isolate a certain part of the audio signal;

– *band-stop filters* or *band-rejection filters:* other names include *band limit filters, bell filters, band-elimination filters* and *T-notch* filters. Unlike the types of filter listed above, these filters reduce a certain frequency band to eliminate undesirable parts of the original signal, such as noise, crackling, etc.

A signal frequency is usually said to be filtered if it is attenuated by 3 dB or more. To distinguish between different types of filters, we can compare their *"bandforms"*. This divides filters into each of the families listed above. Different filters can also have different *"orders"*. Without going into too much technical detail, the order of a filter is determined by the *slope rate* of its attenuation. A first-order filter has a slope of 20 dB/decade, i.e. 6 dB per octave, a second-order filter has a slope of 40 dB/decade, a third-order filter has a slope of 60 dB/decade, and so on.

For band-pass and band-stop filters, the sum of the absolute values of both slopes is calculated.

When applying corrective equalization, we need to remember that we are necessarily introducing phase rotations (shifts) into the signal as a side effect. This can lead to problems for stereo sound recording.

The gain margin plot, known as a Bode plot[1], consists of two subplots corresponding to the *modulus* (gain) and the *argument* (phase), which together allow us to determine the bandform and order of a filter. To calculate this plot as a function of the frequency f in hertz (linear scale for the gain, but logarithmic scale for the frequency), we use the following equation:

$$G_{dB} = 20 log\left(\frac{V_{out}}{V_{in}}\right)$$

where:

– G_{dB}: gain in dB;

– V_{out}: output voltage;

– V_{in}: input voltage;

– V_{out}/V_{in} is called the *"transmittance"* of the filter.

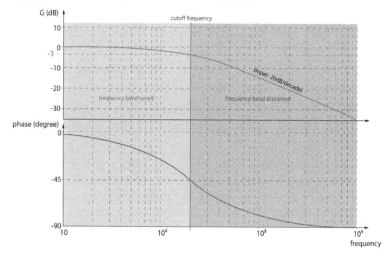

Figure 4.1. *Bode plot of a first-order low-pass filter. For a color version of this figure, see www.iste.co.uk/reveillac/soundeffects.zip*

---

1 Hendrik Wade Bode (1905–1982), American engineer and research who specialized in telecommunications.

By analyzing the eliminated frequencies (the frequencies for which G_{dB} is less than –3 dB), we can determine the bandform of the filter, and by considering the slope, we can find its order.

Equalizers are often used in tandem with spectrum analyzers, which allows the corrections to be more precisely visualized in real time.

Today, digital and software equalizers have become much more precise, and analyzers are easier to find, often integrated directly into Digital Audio Workstations and other audio processing software.

REMARKS.– You might also come across *bell* filters or *Baxandall*[2] filters. Bell filters act on a specific target frequency. Their name comes from the shape of their attenuation curve, which looks like a bell. Baxandall filters are named after their inventor. They are precursors of pitch correctors that were used long before the first equalizers were invented. They can be used to correct bass and treble frequencies, and generally operate at around 1 kHz. They are characterized by the fact that they preserve certain ratios between harmonics. Some sources on equalization also mention the concept of "*Q factor*". This simply refers to the width of the frequency band on which the filter acts.

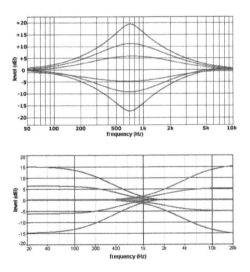

Figure 4.2. *Equalization curves of a bell correction and a Baxandall correction. For a color version of this figure, see www.iste.co.uk/reveillac/soundeffects.zip*

---

2 Peter J. Baxandall (1921–1995) British electronics and sound engineer. In 1952, he invented the pitch correction filter that bears his name. The first Baxandall filters were built from electronic tubes.

4.2. Equalization

The purpose of equalization is to correct the timbre of a sound by boosting or cutting one or more frequency bands in the audio signal.

This type of audio processing is widely used during recording, production and in sound reinforcement systems. It can be implemented using various technological tools, known as *equalizers* or *EQs*, which work in a variety of different ways.

Over time, equalization has become vital to sound engineers, who need it to achieve the right frequency balance in the signal. This allows them to properly separate and showcase each instrument in the mix. For example, the sound of a bass guitar needs to be clearly audible, and should not be allowed to conflict with the bass drum.

We can use different frequency scales to describe the frequencies specific to each type of instrument.

4.2.1. *Frequency bands and ranges*

Frequencies can be divided into six families according to their frequency band:

– treble: from 4 to 20 kHz;

– *high-midrange*: from 2 to 4 kHz;

– midrange: 500 Hz–2 kHz:

– low-midrange: 250–500 Hz;

– bass: 63–250 Hz;

– *sub-bass*: 20–63 Hz.

Each octave (in the musical sense of the term) can be associated with one of these frequency bands. Table 4.1 shows their distribution.

The "register" row in Table 4.1 lists a few technical terms often used to describe the subjective impression created by the sounds in each band.

Octave	C0	C1	C2	C3	C4	C5	C6	C7	C8	C9	C10*
Frequency (Hz) from C to B	16.3 to 30.8	32.7 to 62	65 to 123	131 to 247	262 to 494	523 to 988	1,046.5 to 1,975	2,093 to 3,951	4,186 to 7,902	8,372 to 15,804	16,744 to 19,912
Register	Rumble		Body	Warmth	Low mids	High mids		Presence or attack	Brilliance or clarity		Air

Table 4.1. *Equivalence between octaves and frequencies (* from C to E)*

Figure 4.3 displays the approximate range (frequency spreads) of a few common instruments compared to a standard keyboard.

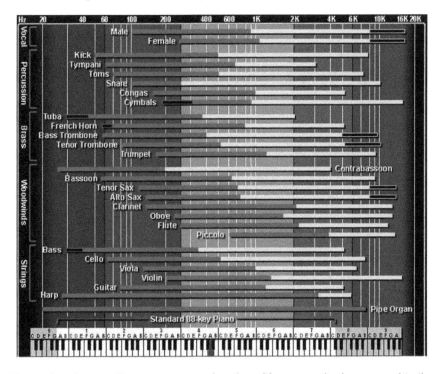

Figure 4.3. *Range of instruments as a function of frequency (top) compared to the notes on a keyboard (bottom) (source: IRN³ (www.independentrecording.net)). For a color version of this figure, see www.iste.co.uk/reveillac/soundeffects.zip*

---

3 We can find an interactive version of this table with plenty of extra information at: www.independentrecording.net/irn/resources/freqchart/main_display.htm.

Having an idea of the range of each instrument will help you to identify the frequency band of the sound you are trying to equalize or modify.

But bear in mind that range is not the only factor. The sound of each instrument has a rich spectrum of harmonics that also need to be taken into consideration.

4.2.2. *Types of equalizer*

There are several kinds of equalizer:

– graphic equalizers;

– parametric equalizers;

– semiparametric equalizers;

– band-stop equalizers or *notch* equalizers;

– linear phase equalizers;

– dynamic equalizers.

4.2.2.1. *Graphic equalizers*

These equalizers are named after their physical design. They are arranged as a series of linear potentiometers (sliders) that cover the entire audible sound spectrum, one for each frequency band, positioned in increasing order from left to right.

One major advantage of graphic equalizers is that they make it very easy to visualize the currently applicable corrections to the incoming signal. The gain/attenuation on each band ranges between ± 18 dB.

Over the years, and as technology has progressed, the number of bands has increased from 5 to 31 (1/3 octave), and in some cases even 64.

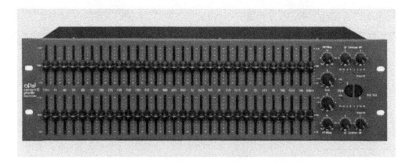

Figure 4.4. *Opal FCS-966, a 31-band graphic equalizer by BSS Audio*

4.2.2.2. Parametric equalizers

Parametric equalizers allow the user to specify the type of filter, cut-off frequency or center frequency, bandwidth and gain.

They are more precise than graphic equalizers, but are more delicate to operate.

Figure 4.5. *Orban 622B, a 2 × 4-band parametric equalizer*

This type of equalizer only acts on a single frequency band at a time, which means that multiple equalizers are required for more elaborate corrections. Most manufacturers offer parametric equalizers with two or four bands.

You can also find so-called passive parametric equalizers, which only reduce the gain, and cannot boost it. A separate active circuit can be used to restore the gain level of the output. The advantage of this type of equalizer is that the signal is largely left unaltered, since the unpowered passive components only create very slight losses.

Figure 4.6. *The well-known passive equalizer EQP-1A by Pultec*

4.2.2.3. Semiparametric equalizers

Semiparametric equalizers combine some of the characteristics of graphic and parametric equalizers.

They typically consist of three (treble, mid and bass) or four frequency bands on which central cut-off frequencies can be defined and the gain can be specified.

Figure 4.7. *The Aphex EQF-2 equalizer and a more recent modernized version, the Aphex EQF-500 model*

4.2.2.4. Band-stop equalizers

Band-stop equalizers are not strictly speaking a separate class of equalizer, but rather a specific feature offered by many equalizers, especially software equalizers.

Band-stop equalization is the reverse of band-pass filtering used to eliminate very narrow frequency bands. For example, a hum at 50 Hz can be eliminated without affecting the surrounding frequencies.

This type of equalizer can be found on some mixing desks and is often used to avoid feedback effects.

Figure 4.8. *An example of a band-stop (notch) filter on a sound reinforcement system*

The filters used by these equalizers are also known as notch filters or rejection filters.

4.2.2.5. Linear phase equalizers

As their name suggests, linear phase equalizers are specially designed to avoid introducing phase shift during processing.

This is an important advantage for the sound engineers responsible for mastering. These equalizers allow phase rotation issues to be avoided during equalization, regardless of the amplification or attenuation of each processed frequency.

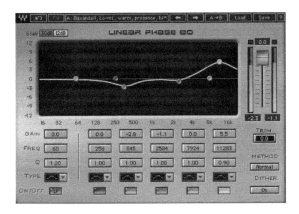

Figure 4.9. *"Linear Phase EQ", a linear phase software equalizer by Waves*

4.2.2.6. Dynamic equalizers

This equalizer is extremely rare in hardware form, and has only become widely available since the introduction of software plugin equalizers.

Dynamic equalizers have filters whose settings change and adapt to the levels of the input signal, or the levels of an external signal if the equalizer includes a *sidechain* function.

Figure 4.10. *The parametric equalizer "Lil Freq EL-Q" by Empirical Labs. This equalizer includes a dynamic equalization feature*

These equalizers work similarly to a compressor (see sections 7.2 and 7.2.2.5), allowing different parameters to be defined for two different filter states. The first state, called the *initial state*, applies whenever the input level is below a specific threshold. The second, often called the *target state*, applies above this threshold.

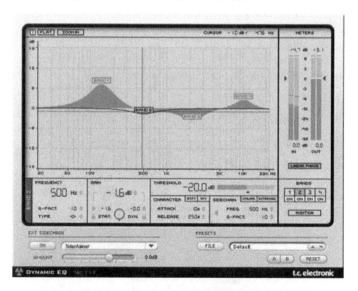

Figure 4.11. *The dynamic software equalizer "Dynamic EQ PowerCore" by TC Electronic*

4.2.3. Examples of equalizers

Table 4.2 lists a few examples of equalizers by type: pedals, racks and software.

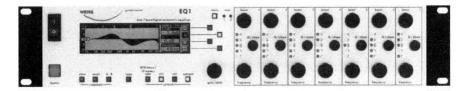

Figure 4.12. *Seven-band, 2-channel digital studio equalizer, EQ1 model by Weiss*

A vast number (hundreds) of equalizer plugins are available on the market. Many software publishers have developed software clones of the most popular studio equalizers.

Type	Manufacturer or publisher	Name or model	Remarks
Pedal	Boss	GE-7	Graphic – 7 bands
	Carl Martin	3 bands – Parametric Pre-Amp	Parametric – 3 bands
	Electro Harmonix	Attack Equalizer	2 bands + mixer
	Electro Harmonix	Knockout	Graphic – 2 bands
	Ibanez	GE9	Graphic – 6 bands
	Maxon	GE601	Graphic – 6 bands
	MXR	M109	Graphic – 6 bands
	MXR	M124	Graphic – 2 × 15 bands
	TC Electronic	Dual Parametric EQ	Parametric – 2 bands
Rack	Apex Audio	DBQ Zero	Graphic – 2 × 30 bands – ± 15 dB
	Avalon	AD 2055	Parametric – 3 bands – 2 channels
	Boss	GE-215	Graphic – 2 × 15 bands – ± 15 dB
	BSS Audio	FCS-966 Opal	Graphic – 2 × 30 bands – ± 15 dB
	DBX	2231	Graphic – 2 × 31 bands – ± 15 dB
	D.W. Fearn	VT-4	Parametric – 1 channel – Electronic tubes
	Empirical Labs	Lil FreQ EL-Q	Parametric – 4 bands – 2 Baxandall

	GML	Model 8200	Parametric – 5 bands – 2 channels
	Gyraf Audio	Gyratec XIV	Parametric – 5 bands – Passive – Electronic tubes
	Manley	Massive Passive EQ	Parametric – Passive – Electronic tubes – 4 bands – 2 channels
	Moog Music	10 Band Graphic EQ	Graphic – 10 bands
	Moog	3 Band Parametric EQ	Parametric – 3 bands
	Orban	672A	Parametric – 8 bands
	Pultec	MEQ5-SS	Parametric – Electronic tube – 2 boosting bands and 2 cutting bands – 300–5,000 Hz
	Rupert Neve Designs	Portico 5033	Parametric – 3 bands + 2 shelf bands[4]
	Samson technologies	E62i	Graphic – 2 × 31 bands – ± 12 dB
	Urei/JBL	527A	Graphic – 27 bands – ± 10 dB
	Summit	EQP-200B	Parametric – 2 bands + 1 low shelf band – Electronic tubes
	Tube-Tech	EQ 1AM	Parametric – 3 band – ± 20 dB – Electronic tubes
	Tube-Tech	PE1C	Parametric – 2 bands – Passive, with electronic tubes
	Weiss	EQ1	Parametric – Digital – 7 bands – 2 channels
Software	Abbey Road	TG Mastering Pack	Parametric – 4 bands + filters Windows: VST OSX: VST; AU
	Analog Obsession	EAQ	Emulation of EAR 825Q Parametric Windows: VST2 OSX: VST2; AU; AAX
	Bomb Factory	Pultec Bundle	Emulation of Pultec EQP-1A, EQH-2, and MEQ-5 Windows: TDM; RTAS OSX: TDM; RTAS; AU

---

4 A *shelf* is a filter that extends from one end of the audible frequency spectrum (or the frequency spectrum modified by the equalizer) up/down to some specific frequency. There are two shelves, high and low, one at each end of the spectrum.

FabFilter	Pro-Q2	24 bands + various filters + (rejection filters, shelves, etc.) Windows: VST OSX: VST; AU
IK Multimedia	T-Racks Classic Equalizer	Parametric – 6 bands + filters Windows: VST; RTAS; AAX OSX: VST; RTAS; AAX
MeldaProduction	MEqualizerLinearPhase	3 equalization algorithms – 9 filters Windows: VST; VST3; AAX OSX: VST; VST3; AAX
Overtone DSP	EQ4000	Parametric Windows: VST; VST3; AAX OSX: VST; VST3; AAX; AU
PSP Audioware	Retro Q	Parametric – 3 bands – Vintage EQ emulation Windows: VST; AAX OSX: VST; AAX; AU
RJProjects	Aqualizer	Freeware – Graphic – 32 bands – ± 12 dB Windows – OSX: VST
Rob Papen	RP-EQ	8 bands + filters + spectrogram Windows: VST; AAX OSX: VST; AAX; AU
Softube	Tube-tech ME1B	Emulation of Tube-Tech ME-1B based on Pultec MEQ-5
Softube	Tube-Tech PE1C	Emulation of Tube-Tech PE-1C
Sonnox	Oxford EQ	Parametric – 5 bands Windows: TDM; RTAS; AAX OSX: TDM; RTAS; AAX
SPL	Free Ranger	Freeware – Graphic – 4 bands Windows: VST; VST3; AAX OSX: VST; VST3; AAX; AU
TC Electronic	Assimilator	Based on curves – Presets – Limiter Windows: VST OSX: VST; MAS
Universal Audio	Pultec Passive EQ Collection	Emulation of Pultec EQP-1A, HLF-3C, and MEQ-5
Universal Audio	NEVE 1073 EQ	NEVE 1073 emulation

Voxengo	HarmoniEQ	Parametric – Stereo and 5.1 surround mastering equalizer – Dynamic equalization – Spectrum analyzer Windows: VST; VST3 OSX: VST; VST3; AU
Voxengo	GlissEQ 3	Parametric – Stereo and 5.1 surround equalizer – Spectrum analyzer – Dynamic equalization Windows: VST; VST3 OSX: VST; VST3; AU
Waves	Linear Phase	Mastering equalizer – 9 types of filter Windows: VST; VST3; AAX OSX: VST; VST3; AAX; AU
Waves	Q10	Paragraphic – 10 bands Windows: VST; VST3; AAX OSX: VST; VST3; AAX; AU
Waves	H-EQ	Hybrid (parametric + curves) – 7 types of filter Windows: VST; VST3; AAX OSX: VST; VST3; AAX; AU
Waves	RS56 Passive EQ	Passive parametric – 3 bands – 6 types of filter Windows: VST; VST3; AAX OSX: VST; VST3; AAX; AU
Waves	V-EQ4	Parametric – 4 bands – Emulation of NEVE 1081 console Windows: VST; VST3; AAX OSX: VST; VST3; AAX; AU

Table 4.2. *Examples of each of the three types of equalizer*

4.2.4. *Tips for equalizing a mix*

Many modern mixing desks already have an integrated four-band parametric equalizer, often with a low-cut filter. These tools will probably be sufficient for most things, although they of course cannot compare with the level of sophistication offered by one or several external equalizers.

Filtering Effects   95

Figure 4.13. *Integrated parametric filters on a mixing desk*

The first question is where the equalizer should be placed in the audio processing chain: before or after the compressor (see Chapter 7 – Dynamic effects)? In fact, there is no single answer to this. As a rule of thumb, if you are cutting frequencies, put the compressor first, and if you are boosting frequencies, put the equalizer first.

Be wary of equalizers that add color (non-transparent equalizers) when you boost certain frequencies. This can affect the style of the mix, which may be undesirable. For initial separation and rough filtering, use an equalizer that is as neutral as possible.

In my personal opinion, the best approach is to start by cutting certain frequencies to clean the mix before boosting anything. Work surgically and carefully on narrow bands. You usually will not need to modify the gain by more than ± 3 dB, which is already a lot.

The basic rules described in the following can be useful. Keep in mind which frequency bands are occupied by each of the instruments in your mix.

If your mix is too aggressive, cut the band between 1.2 and 3 kHz. This will soften the sound a little. You can muffle the sound by pushing the gain toward the negatives. However, bear in mind that some instruments and vocals can be affected by this and might lose clarity. To compensate, you can boost the higher part of this band, between 2 and 3 kHz.

If the lower frequencies lack space or air, you can cut the 80–100 Hz region of the mix. Since this is the bottom of the spectrum, you can use a shelf to cut frequencies below 100 Hz. But be careful not to squash the bass. You can make it more distinctive by boosting its specific tone.

If your mix seems cluttered and confused, you can carefully cut the mid band from 100 to 300 Hz.

High frequencies are often the most difficult to equalize, since too much gain will create an imbalance that can disrupt the whole mix, resulting in a sound that quickly becomes uncomfortable.

If you are having problems with excessive *sibilance*, you will get better results with a *de-esser* (see section 7.5) than an equalizer.

If you think that your mix has too much air, you can configure a shelf at the top of the spectrum above 15 kHz and cut slightly. Make sure that you keep some *presence*.

Note that we tend to cut frequencies rather than boost them. You will get the feel for this over time. The most important thing is to keep listening to your mix, constantly playing with the *bypass* to get a clearer picture of the difference with and without your adjustments. Take your time, and allow your ears to rest whenever they need to. Equalization can be very tricky, so do not hesitate to revert any changes that do not work. Do not forget that a single change can affect the entire mix.

Accentuation can be used to shape the sound space of the mix or add color. If you are lacking in color, you can choose an instrument to accentuate using a separate equalizer from the one used to make the adjustments described above, perhaps a more stylized model that specifically fits the musical style of your mix. Working on narrow bands will allow you to increase the presence or brilliance of an instrument, whereas working with wider bands can enhance an ensemble. Remember that the wider the band, the smaller the required gain, usually less than 2 dB.

The sound space can be managed according to a simple principle that reflects how things work physically in reality: the trebles should be less present when the source is further away and vice versa. Play with the gain in small increments of 0.25 dB in both directions until you find the right location for each sound. The higher the frequency (>10 kHz), the lower the required corrections. If you want to expand or broaden the space with respect to one or multiple instruments, one possible solution is to use a shelf to create slight accentuation above 5 or 6 kHz.

If you are using a shelf for the bass frequencies (<300 Hz) of an instrument, you should move it to the edge of the sound field.

These instructions might seem vague, but equalizing mixes is difficult and very subjective. If you are working on a specific instrument, listen carefully, first to the effect of the changes on the instrument itself, and then to their effect on similar instruments, and finally to the effect on the whole mix.

I use this three-step approach myself for my work, it works very well for me. It is one way of doing things, but everyone needs to forge their own path. There are no hard-and-fast rules, and the only things that you need to answer to – it is impossible to emphasize this enough – are your ears.

4.3. Wah-wah

Some people might find it strange to include the *wah-wah* in this chapter. But from the perspective of physics, wah-wah is definitely a filtering effect, so it fully deserves its place here.

Wah-wah is an effect that alters a musical passage to sound like a human voice. Musicians use it as a way of playing more expressively. It can be used with most instruments, but is most popular with brass instruments and electric guitar.

REMARK.– Wah-wah is pronounced like it is written. You might also come across the alternative spellings "wha-wha" or "wa-wa".

4.3.1. *History*

Wah-wah was invented in the mid-1960s. The effect imitates the crying sound that can be made with the human voice. The electronic wah-wah pedal was invented by the engineer Brad Plunkett. He discovered the effect when he was modifying a Vox amplifier in an attempt to make it cheaper to manufacture without sacrificing any sound quality.

The original Vox amplifier had a switch known as an MRB (*Mid-Range Boost*) that increased the presence of the sound by boosting the central part of its spectrum

(300–500 Hz). Plunkett replaced this switch with a potentiometer, which at the time was much less expensive.

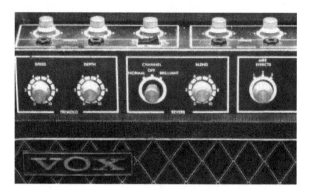

Figure 4.14. *The three-position MRB EFFECTS switch (right) used in the first Vox amplifiers*

He later realized that varying the potentiometer while a guitarist was playing through the amp created this wah-wah effect. He therefore decided to mount this system into a Vox Continental organ pedal so that it could be operated by foot. Thus, the wah-wah pedal came into existence.

In 1967, Vox placed the pedal on the market with two different names: "Cry-Baby", because it sounds like a crying baby, and "Vox Wah-Wah". The electronics in both pedals were identical.

Figure 4.15. *Two versions of the Vox wah-wah pedal, the "Vox Wah-Wah" and the "Cry-Baby"*

The wah-wah effect was already familiar to trumpeters, who had been using it since the 1920s by inserting a mute, known as the "wah-wah", into the bell of the trumpet.

Figure 4.16. *A "wah-wah" mute designed for a trumpet*

This mute simply consisted of a sliding tube and a cup. By sliding the tube back and forth, the trumpeter can modify the sound. By varying how much the tube is blocked while playing, the trumpeter controls the wah-wah effect.

REMARK.– By pressing a regular toilet plunger against the bell of the trumpet and varying the gap, jazz trumpeters can simulate the wah-wah effect.

4.3.2. *Theory*

The wah-wah pedal is based on a selective amplification system (affecting only certain parts of the sound spectrum) whose *central frequency* is modulated.

When the musician tilts the pedal, the boosted and filtered regions of the spectrum slide up or down.

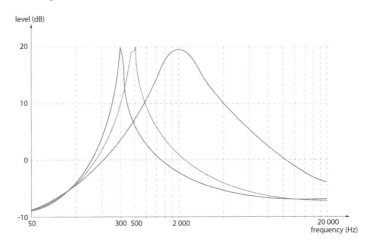

Figure 4.17. *Frequency response of a classic wah-wah pedal. The three curves (red, blue and green) represent the filtering and amplification of each region of the sound spectrum depending on the position of the pedal, from the up position to the down position. The spectrum slides from left to right as the pedal is lowered. For a color version of this figure, see www.iste.co.uk/reveillac/soundeffects.zip*

Electronically, there are several ways of creating this effect. Without going into any detail, we can quote:

– the oldest, used from 1960 to 1975, involved an array of transistors plus a *resonant filter*;

– from 1975 to 1990, operational amplifiers plus a resonant filter were used;

– digital filters have been used since the 1990s.

There are also systems controlled by *Pulse Width Modulation* that can be used to program the wah-wah effect, allowing extra dimensions (stereo rebound, periodic sound variation, acceleration, deceleration, rotation, etc.) to be added.

4.3.3. Auto-wah

This type of pedal, also known as T-wah, Q-wah or an envelope filter, is most commonly used with guitars, much like its close relative, the wah-wah pedal, although it can of course also be used with other instruments.

The fundamental difference with a wah-wah pedal is that the auto-wah effect is controlled by the volume or attack dynamics of the input signal. Some variants use a *low-frequency oscillator* with a configurable frequency to trigger the effect.

One of the most famous models of auto-wah is the Mu-Tron III created by Mike Beigel[5].

Figure 4.18. *The famous Mu-Tron III auto-wah pedal, an early model from the 1970s*

---

5 One of the founders of Musitronics Corporation in the 1970s, together with Aaron Newman. Musitronics Corporation was an American company based in New Jersey that specialized in audio filters and sound effects. In 2013, Mike Beigel founded the Californian company Mu-FX.

4.3.4. *Examples of wah-wah pedals*

Table 4.3 lists a few examples of wah-wah pedals, including the best-known models (Vox, Dunlop, Morley, etc.), and a few auto-wah pedals.

Manufacturer	Name or model	Remarks
Boss	AW-3	Wah-wah – Auto-wah – Humanizer
Carl Martin	2 Wah	Wah-wah pedal
Digitech	Synth Wah	Auto-wah + filters + effects
Dunlop	DCR 2SR Cry Baby	Rack – Wah-wah – 6-band equalizer – Various settings – External control pedal
Dunlop	GCB95	Wah-wah pedal
Dunlop	Cry Baby	Wah-wah pedal
Dunlop	JH1 Jimi Hendrix	Wah-wah pedal
Electro Harmonix	Doctor Q	Auto-wah
Electro Harmonix	Mini Q-Tron	Auto-wah
Electro Harmonix	Doctor Q Nano	Auto-wah
Electro Harmonix	Wailer Wah	Wah-wah pedal
Electro Harmonix	GE9	Graphic – 6 bands
Mooer	Funky Monkey	Auto-wah – 3 modes
Morley	Power Wah	Wah-wah pedal
Morley	Steve Vai Bad Horsie Wah	Wah-wah pedal
Morley	Pro Series Wah	Wah-wah pedal
Mu-Fx	Tru-Tron 3X	Auto-wah – Based on the Mu-Tron filter
MXR	M120 Auto Q	Auto-wah – Various filter settings
MXR	M124	Graphic – 2 × 15 bands
Vox	V847	Wah-wah pedal
Vox	King Wah	Wah-wah pedal
WMD	Super Fatman	Auto-wah – 12 filters – Various settings

Table 4.3. *Examples of auto-wah and wah-wah pedals*

4.4. Crossover

There is one final effect that I want to mention in this chapter. Crossover is an effect that is mostly used in studios, although it is also used for sound reinforcement.

Crossover is a filter that divides an audio signal into two or more frequency ranges. Crossover filters are often passively integrated into speakers with two or three tracks, enabling the speakers to cover the full spectrum (bass, mid, treble).

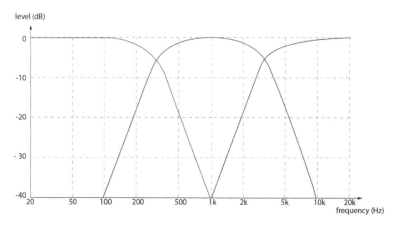

Figure 4.19. *The three typical curves of a 3-way crossover (blue: bass; green: mid; red: treble), showing the gain as a function of the frequency. For a color version of this figure, see www.iste.co.uk/reveillac/soundeffects.zip*

You can also find them in systems with subwoofers, such as home theater sound systems. Finally, they are used by some specialist sound processing studios.

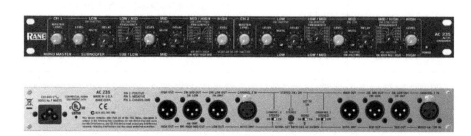

Figure 4.20. *Front (top) and rear (bottom) of an active 2- or 3-way crossover, the AC 23S by Rane. At the rear, we can see one input and three outputs for each of the two channels*

When the crossover is turned on, it splits the audio signal before the final amplification stage. Crossover is often combined with a delay, an equalizer, or a limiter (surround systems).

Many racks have integrated active crossovers, but crossovers can also be found in the form of preprocessing plugins. Table 4.4 lists a few examples.

Type	Brand or publisher	Name or model	Remarks
Rack	ART	311	Active – 2-way stereo + mono sub
	Behringer	CX-2310	Active – 2-way stereo – 3-way mono
	Behringer	CX-3400	Active – 2- or 3-way stereo – 4-way mono – Limiter
	DBX	234XS	Active – 3-way stereo – 4-way mono – Phase inverter
	Lab Audio	AXC-23 XL	Active – 2- or 3-way stereo
	LD Systems	X223	Active – 2-way stereo – 3-way mono
	Peavey	PW350XO	Active – 2- or 3-way stereo – 4- or 5-way mono
	Phonic	PCR2213	Active – 2-way stereo – 3-way mono – Delay (0-4 ms)
	Rane	MX 22	Active – Stereo – 2 tracks
	Rane	AC 23S	Active – 2- or 3-way stereo – 4- or 5-way mono – Delay
	Samson	S-3	Active – 2- or 3-way stereo – 4-way mono + sub – Delay – Limiter
Software	Hornet	3xOver	3 bands Windows – OSX: VST2; VST3
	RS-Met	CrossOver	Freeware – 4 bands Windows – OSX: VST

Table 4.4. *Examples of crossovers*

4.5. Conclusion

As you can see, there is a vast diversity in the nature and applications of filtering effects, which are used for purposes ranging from audio signal equalization to guitar effects and sound reproduction.

Equalization is an essential part of mixing or mastering, and is one of the key factors in sound quality and rendering. However, it can be a difficult skill to perfect.

The wah-wah effect is now part of the standard repertoire of every guitarist. Others also use it with bass guitar, electric pianos or keyboards, or even *clavinet*[6] electric clavichords.

Crossover filters are primarily used within multichannel systems, as well as for certain specialized types of audio processing.

---

[6] Widely used electric clavichord manufactured by Hohner.

5

Modulation Effects

Modulation effects introduce a slight delay into the audio signal, then readd the signal thus obtained to the original signal.

The main types of effect in this category are:

– *flanger*;

– *phaser*;

– *chorus*;

– *rotary*, *univibe* or *rotovibe*;

– *ring modulation*.

5.1. Flanger

Flanger is the oldest and also the simplest from a technical point of view, so we shall discuss it first. In a certain sense, the other modulation effects, phaser and chorus are just more or less sophisticated derivatives of the flanger effect.

Flanger creates a sound like a jet engine. Although it can be applied to any audio signal, it is most commonly used with guitars and electric keyboards, giving them a very specific and recognizable color.

5.1.1. *History*

It is not completely clear where flanger came from, but there are a few major milestones that we can quote:

– Les Paul[1] may have discovered it in the 1940s while tinkering with the machines in his studio;

– David S. Gold and Stan Ross from Gold Star Studios in Hollywood claim to have used it in their 1959 release of the song "The Big Hurt" by Toni Fisher;

– in 1966, Ken Townsend, a sound engineer at Abbey Road Studios (formerly EMI), is probably the person who played the largest part in popularizing the effect. In order to be able to overlay duplicates of John Lennon's voice without having to record it twice, he invented a technique known as *"Artificial Double Tracking (ADT)"*. His technique worked, but also created a rotational effect whenever the scrolling speed of one of the recorders was accidentally changed, causing flanging[2] (buckling) in the tape;

– George Chkiantz, a sound engineer at Olympic Studios, used it for the 1967 recording "Itchycoo Park" by Small Faces;

– the first stereo version of the flanger may have been created by the producer and sound engineer Eddie Kramer for the song "Bold as Love" by Jimi Hendrix, released in 1967;

– in 1968, the producer and studio director Warren Kendrick invented a reliable way of controlling the effect by using two Ampex recorders and a screwdriver placed between them.

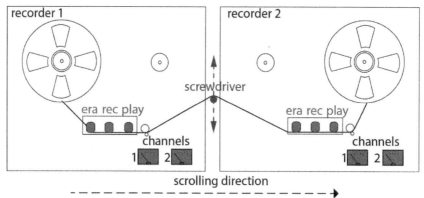

Figure 5.1. *The method invented by W. Kendrick for creating the flanger effect. The position of the screwdriver controls the extent of the flanging (buckling) of the tape, allowing the recording on one track to be shifted relative to the other. In the high position, the two recordings are further apart; in the low position, they are closer together. For a color version of this figure, see www.iste.co.uk/reveillac/soundeffects.zip*

---

1 Full name Lester William Polsfuss, 1915–2009, American guitarist and inventor.
2 Lateral deformation of a long component under normal compression.

The first analog flangers appeared in the 1970s, as musical electronics began to become available, because of new components such as integrated circuits.

They would soon be followed by digital flangers based on dedicated digital signal processors.

Today, a variety of flanger software plugins are also available.

5.1.2. *Theory and parameters*

A very short delay, between 5 and 50 ms, is added to the audio signal. The delayed signal thus obtained is then rerouted to the input via a connection called the *feedback*. The delay time is determined by a low-frequency oscillator (LFO). A mixer then combines the original (*dry*) signal with the processed (*wet*) signal.

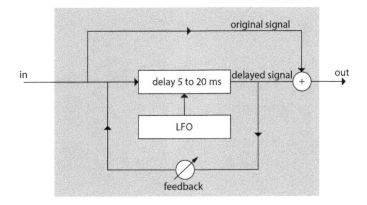

Figure 5.2. *Flowchart of the flanger effect*

The physics of the effect can be described as shown in Figure 5.3.

For example, if a signal with a frequency of 1 kHz is added to another signal of same frequency with a delay of 1 ms (a multiple of this frequency), the periods of the signals line up (two signals are in phase) and the amplitude increases.

However, if the delay is 1 ms and the frequency is 500 Hz, the periods of both signals cancel (the two signals are out of phase by 180°).

In reality, sounds are much more complex and contain many signals with different frequencies. Different parts of the sound will therefore be modified differently. Some frequencies will be boosted, some will be cut and some will cancel out completely. The same is true for the harmonics.

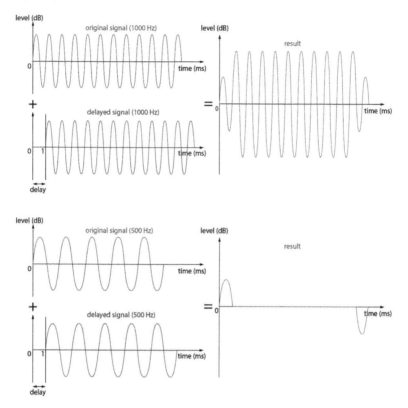

Figure 5.3. *The influence of the delay time as a function of the frequency when summing sine waves. For a color version of this figure, see www.iste.co.uk/reveillac/soundeffects.zip*

By referring to Figure 5.2, we can identify each of the parameters that you might encounter when configuring a flanger.

The most important parameters are:

– the *delay time*: this defines the time difference between the original signal and the processed signal. Usually expressed in ms;

– the *speed* (or *rate*): this parameter defines the modulation speed. Usually expressed in Hz;

– the *mix* (dry/wet): the mix is the ratio of original signal to processed signal;

REMARK.— The mix is often called the *depth* or *width*, especially on some models of pedal. This is a separate concept from the depth parameter defined below.

– the *feedback* (or *regen*): this parameter controls the quantity of processed signal returned to the input. Usually expressed as a percentage between 0 and 100%.

Figure 5.4. *Three flanger pedals: "BF-3" by Boss, "Vortex" by TC Electronic and "Micro Flanger" by MXR*

Some flangers also include the following additional parameters (the list is far from exhaustive):

– the *depth* (occasionally *width*): the depth defines the extent of the effect and is usually expressed as a percentage between 0 and 100%. At 0%, the delay is a fixed constant value; at 50%, the delay varies over a time interval equal to half of its value; at 100%, the delay varies over its full value. For example, if the delay is 5 ms, then 0% depth gives a fixed delay of 5 ms; 50% depth gives a delay varying between 0 and 2.5 ms; and 100% gives a delay varying between 0 and 5 ms;

– the *waveform*: this defines the modulation shape, which can be a sine wave, a triangle wave, etc.;

– the *stereo offset*: this distributes the flanging effect over the two channels of a stereo signal and is usually expressed in degrees between 0 and 180°. At 0°, the effect is in phase on both channels; at 180°, the flanging effect is minimal on the left channel and maximal on the right channel; and so on;

– the *output level* (gain): this parameter defines the overall output volume (original signal + processed signal) and usually expressed in dB;

– the *filter*: this defines a threshold between two frequency bands: *low* and *high*. It is usually expressed in Hz or kHz.

Figure 5.5. *The "MetaFlanger" plugin by Waves, with its various settings*

5.1.3. *Models of flanger*

In Table 5.1, you can find a few examples of flanger pedals, racks and plugins.

Some plugins are part of a larger bundle of effects.

Type	Manufacturer	Name or model	Remarks
Pedal	Alexander Pedals	F.13	Three types of flanger
	Boss	BF-3	Flanger – Guitar or bass input – Tap tempo – Mono or stereo output
	Electro Harmonix	Electric Mistress	Multieffects: Flanger – Filter matrix
	Moog	MF Flanger	Analog – 2 types of flanger
	MXR	M117R	Flanger for any instrument
	MXR	M152 Micro Flanger	Analog flanger
	TC Electronic	Vortex	Analog flanger – Stereo inputs/outputs
Rack	Avid	Eleven Rack	Multieffects – Guitar amp simulators
	Digitech	GSP1101	Multieffects – Guitar amp simulators
	Ibanez	AD-220	Analog – Delay and flanger (vintage 1970)

	Lexicon	MX Series	Digital multieffects
		MPX Series	Digital multieffects
	Line 6	Pod Pro	Multieffects – Guitar amp simulators
	MXR	Model 126	Flanger and doubler (vintage)
	TC-Electronic	D-Two	Digital multieffects – Multitap
		M Series	Digital multieffects
Plugins	Auddifex	STA Flanger	Windows: VST; VST3; RTAS OSX: VST; VST3; AU; RTAS
	Blue Cat Audio	Flanger	(freeware) – Windows: VST; AAX; RTAS; DX OSX: VST; AAX; RTAS; AU
	Blue Cat Audio	Stereo Flanger	(freeware) – Windows: VST; AAX; RTAS OSX: VST; AAX; RTAS; AU
	Eventide	Instant flanger	Windows: VST; AAX OSX: VST; AU; AAX
	Kjaerhus Audio	Golden modulator – GMO-1	Multieffects – Windows: VST
	Nomad factory	Blue Tube – Analog chorus/flanger	Windows – OSX: VST; RTAS
	Softube	Fix flanger	Windows: VST; VST3; AAX; OSX: VST; VST3; AU; AAX
	Waves	MetaFlanger	Windows: VST; VST3; AAX; RTAS OSX: VST; VST3; AAX; AU; RTAS

Table 5.1. *Examples of flangers*

5.2. Phaser

Phaser is a slightly more complex effect derived from the flanger. The input signal is divided into several frequency bands using multiple band-pass filters. Then, a delay is applied to each of them, as shown in Figure 5.6.

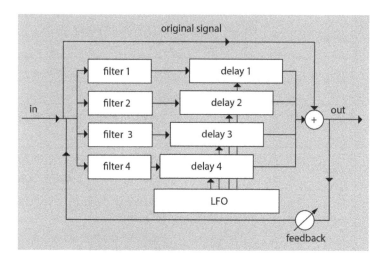

Figure 5.6. *Flowchart of the phaser effect*

The number of bands is not fixed, but there are usually at least four. Like the flanger, the delay on each band is very short (<20 ms).

Phasers can be viewed as a combination of multiple flangers.

Figure 5.7. *Three phaser pedals: Moog MF-103, Boss PH-1 and MXR Phase 100*

Modern phasers can have over a dozen[3] filters and frequency bands. Some have special features that only act on certain parts of the sound spectrum.

Figure 5.8. *The "Blue Cat's Phaser" plugin*

5.2.1. *Examples of phasers*

The rack section of Table 5.2 overlaps with the list of flangers cited earlier, since racks often offer multiple effects.

Some plugins are part of a larger bundle of effects.

Type	Manufacturer	Name or model	Remarks
Pedal	Boss	PH-1	Vintage: 1977-1981 – Analog 4- stage
	Boss	PH-2	Vintage: 1988-2001 – Analog 12-stage
	Electro Harmonix	Bad stone	2015 – 6-stage
	Ibanez	PT9	Vintage, 1980s
	Moog	MF-103	Analog – 6- or 12-stage
	Morley	Pro phaser	Phaser + Expression pedal
	MXR	M101 – Phase 90	From 1970 to today (reissue) – One of the classics – Analog
	MXR	M107 – Phase 100	From 1970 to today (reissue) – Analog – Modified version of the Phase 90
	MXR	M290	Combines the Phase 45 and the Phase 90
	TC Electronic	Blood moon phaser	Vintage – 3 settings: rate, depth, feedback
	TC Electronic	TonePrint Helix Phaser	Four settings: speed, depth, feedback, mix
	T-Rex engineering	Tremonti phaser	Four- or 8-stage – four settings: bite, level, depth, rate

---

3 The number of filters is commonly called the number of stages: 4-stage, 12-stage, etc.

Type	Brand	Models	Remarks
Rack	Avid	Eleven Rack	Multieffects – Guitar amp simulators
	Digitech	GSP1101	Multieffects – Guitar amp simulators
	Lexicon	MX Series	Digital multieffects
		MPX Series	Digital multieffects
	Moog	Moog music 12-stage phaser MKPH	4-, 6-, 8-, 10- and 12-stage – More advanced rack version of the MF 103 pedal
	TC-Electronic	D-Two	Digital multieffects – Multi-tap
		M Series	Digital multieffects
Plugins	Audiffex	STA Phaser	Windows: VST; VST3; RTAS OSX: VST; VST3; AU; RTAS
	Audio Damage	Phase two	Emulation of Mutron BiPhase – Windows: VST OSX: VST; AU
	Blue Cat Audio	Phaser	(freeware) – Windows: VST; VST3; AAX; RTAS; DX (freeware) – OSX: VST; VST3; AU; AAX; RTAS
	Eventide	Instant phaser (Anthology X)	Windows: AAX – VST OSX: AAX – VST – AU
	Kjaerhus Audio	Classic phaser	(freeware) – Windows: VST;
	Kresearch	KR phaser	(freeware) – Windows: VST; VST3; AAX (freeware) – OSX: VST; VST3; AU; AAX
	MeldaProduction	MPhaser	(freeware) – Windows: VST (freeware) – OSX: VST; AU
	Nomad Factory	Free phaser	(freeware) – Windows: VST
	Nomad Factory	Blue tubes analog phaser APH-2S	Windows: VST; AAX; RTAS OSX: VST; AU; AAX; RTAS
	Smart Electronix	SupaPhaser	(freeware) – Windows: VST (freeware) – OSX: VST; AU
	SoundToys	Phase mistress	Windows: VST; AAX OSX: VST; AU; AAX
	Synthescience	Phaser 2/8	(freeware) – Windows: VST

Table 5.2. *Examples of phasers*

5.3. Chorus

Chorus works according to a similar principle to the flanger and phaser effects. The fundamental difference is that chorus uses higher delays of between 20 and 80 ms.

The goal is to create the impression of multiple instruments or voices. In reality, whenever several people play together, there are small differences between each of them – they are never perfectly synchronized. The chorus recreates this effect.

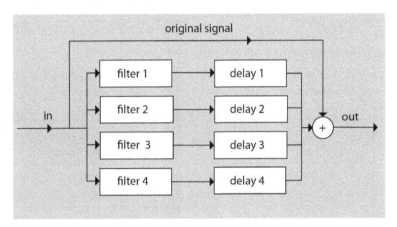

Figure 5.9. *Flowchart of the chorus effect*

A slight pitch (tuning) difference can be applied to each filter to simulate the fact that multiple instruments are always very slightly out of tune with each other.

Figure 5.10. *Three chorus pedals: "Small Clone" by Electro Harmonix, "Stereo Chorus" by MXR and "Corona Chorus" by TC Electronic*

Not every instrument has exactly the same range, so filtering affects different instruments differently. Thus, by varying the volume of each filter, we can create the impression of differentiation between instruments by simulating their distance from the sound recording equipment.

5.3.1. *Examples of chorus*

Again, the rack section of Table 5.3 overlaps with the flanger, since each rack can have multiple effects.

Some plugins are part of a larger bundle of effects.

Type	Manufacturer or publisher	Model	Remarks
Pedal	Boss	CE-20	Released in 2001 – 6 types of chorus – 4 presets – Various settings
	Boss	CH-1	Available since 1989 – Analog until 2001 – Digital today
	Boss	CE-5 chorus ensemble	Available since 1991 – Analog until 2001 – Digital today – 2 filters
	Electro Harmonix	Small clone MK2	Analog – Reissue from the 1990s
	Electro Harmonix	Neo clone	Released in 2010 – Analog – Simple settings
	Ibanez	TC10	Released in 1986 – Analog – 2 types of chorus
	Korg	NuVibe	Chorus and vibrato – Reissue of the famous "Uni-vibe" – Allows the waveform to be configured
	Moog	MF Chorus	Analog – 3 types of chorus
	Providence	ADC-4 Amandine chorus	Analog – 3 types of chorus
	MXR	M234	Analog – 2 filters

	MXR	M134	Various settings – Bass filter
	TC Electronic	Corona chorus	Three types of chorus – Toneprint – USB port
	TC Electronic	Corona mini	Simple settings – Toneprint – USB port
	Strymon	Ola	Three types of chorus – 3 types of dynamics – Various settings
Rack	Avid	Eleven Rack	Multieffects – Guitar amp simulators
	Digitech	GSP1101	Multieffects – Guitar amp simulators
	Lexicon	MX Series	Digital multieffects
		MPX Series	Digital multieffects
	Moog	Moog music 12-stage phaser MKPH	4-, 6-, 8-, 10- and 12-stage – More advanced rack version of the MF 103 pedal
	TC-Electronic	D-Two	Digital multieffects – Multi-tap
		M Series	Digital multieffects
Plugins	Kjaerhus Audio	Classic phaser	(freeware) – Windows: VST
	Audiffex	STA chorus	Windows: AAX; RTAS; VST2; VST3 OSX: AAX; AU; RTAS; VST2; VST3

Table 5.3. *Examples of chorus*

REMARK.– The "CE-1" model by Boss, released in 1976, was one of the first ever chorus pedals. But it was actually designed to recreate an effect from a legendary amp, the "JC-120" by Roland[4], which stands for "Jazz Chorus 120". As its name suggests, the amp had an integrated chorus effect.

5.4. Rotary, univibe or rotovibe

This effect simulates the sound of *Leslie rotating speakers*, an amplification system specifically designed for electronic organs including Hammond organs.

---

4 Boss is a division of the Japanese company Roland.

Figure 5.11. *Exterior and interior view of a Leslie speaker, 122 model*

These speakers are named after their inventor, Donald Leslie[5].

5.4.1. *History*

After acquiring a Hammond organ together with a so-called "tone cabinet" amplification system in the 1930s, Donald Leslie began to look for an alternative system, finding that the sound was flat and lacked warmth. He wanted to recreate the sound effects of a pipe organ, resulting in the development and creation of his first rotating speaker in 1937.

He presented his invention to Laurens Hammond, but Hammond was not pleased, anticipating that it would compete with his own amplification system.

Spurred on by Hammond's rejection, D. Leslie created his own company, Electro Music Inc., which marketed its first Leslie speaker in 1940 under the name "Vibratone – 30A model".

These speakers were then improved over time. In some models, the tube amps (20–50 W) were replaced by more powerful transistor amps (90 W and above). In 1963, in addition to the standard speed of rotation, which was called "tremolo", they

---

5 Donald James Leslie, 1911–2004, American inventor. Member of the American Music Conference Hall of Fame, alongside Laurens Hammond (who invented the Hammond organ) and Leo Fender (who founded Fender).

were fitted with a second, slower speed, called "chorale". Spring reverb was added to some models (Hammond – see section 8.1.2).

Figure 5.12. *Two generations of Leslie speakers: The first model – 30 A (1940) with tubes, and the 760 model with transistors (solid-state – 1970s)*

Special control panels and preamplifier pedals, known as "combo pedals", were invented to connect the speakers to any type of organ, or even other instruments such as guitars.

Figure 5.13. *A "Combo" preamplifier pedal manufactured by Leslie*

In the 1990s, Electro Music Inc. was acquired by Hammond-Suzuki.

Leslie speakers still exist today, and new speakers are still available on the market, manufactured and distributed by Hammond-Suzuki.

5.4.2. *Theoretical principles*

The idea is to send the low frequencies of an amplified sound signal to a rotating drum with an inclined, rounded plane, and to send the high frequencies to a rotating horn, as shown in Figure 5.14.

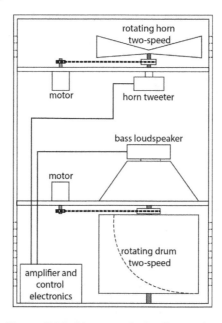

Figure 5.14. *Diagram of a Leslie speaker*

The rotation varies the sound modulation, making it seem to move back and forth.

This phenomenon is caused by the Doppler effect, which was discovered by the physicist Christian Doppler[6] in 1845. The Doppler effect arises from the receiver's perception of the offset between the emitted wave and the received wave whenever the emitter and the receiver are moving relative to each other.

---

6 Johann Christian Doppler, Austrian physicist, 1803–1853.

I am sure you have noted the sound made by a passing car when you are standing at the edge of the road. The sound is higher pitched when the car is moving closer, and lower pitched when it is moving away.

Whenever a sound source moves toward an observer, the sound waves are compressed, reducing the wavelength λ_2 and increasing the frequency of the sound. The perceived sound is therefore higher. Whenever the source moves away from the observer, the opposite phenomenon occurs. The sound waves dilate, the wavelength λ_1 increases and the frequency decreases, making the sound appear lower.

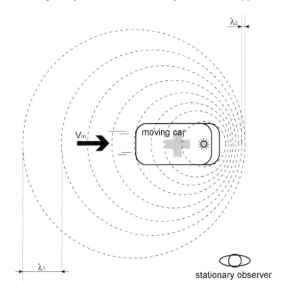

Figure 5.15. *The Doppler effect*

$$\lambda_1 = (c + V_m)T$$

$$\lambda_2 = (c - V_m)T$$

where:

– c: speed of sound (340 m/s);
– λ_1: wavelength 1 (m);
– λ_2: wavelength 2 (m);
– V_m: speed of the moving object (m/s);
– T: period (s).

One simple way to visualize the Doppler effect is to imagine yourself standing in the water at the seaside. Suppose that the waves hit your feet every 5 s. If you run further into the sea, the waves will hit you more often as you run, and so their frequency is higher (compression). If you turn around and run back to shore, the waves will hit you less often as you run, and so their frequency is lower (dilation). Note however that moving around in the water does not change the total number of waves.

5.4.3. Leslie speakers

Many brands, such as Elka, Sharma, Dynacord, Echolette, Fender, Solton and others, have created their own copies of Leslie speakers.

Figure 5.16. *Leslie speaker clones: Sharma, Echolette, Dynacord and Elka*

Table 5.4 lists a few models of speaker that are still available today.

Manufacturer	Model	Remarks
Motion Sound	KBR-3D	100 W + 45 W (12AX7 pre-amp tubes) for the rotary channel – Stereo – Equalizer
Motion Sound	PRO 145	130 W bass and 70 W treble (12AX7 pre-amp tubes)
Motion Sound	1771 – 1771W	200 W + 2 rotors
Motion Sound	PRO-3X	45 W – Equalizer – Very compact
Hammond Leslie	122XB	40 W with tubes – 2 rotating systems – Reissue of the 122 Leslie model
Hammond Leslie	3300	220 W bass + 80 W treble – Pre-amp with tubes – Solid-state amp – 2 rotating systems
Hammond Leslie	2101	50W rotating horn – 4 fixed 75 W speakers – Overdrive with tubes

Table 5.4. *Modern speakers available for purchase*[7]

5.4.4. Examples of rotary or univibe pedals

Figure 5.17. *Four rotary effects pedals: "Rototron" by Pigtronix, "Leslie Digital" and "RT-20" by Boss, and "Lex" by Strymon*

---

7 These speakers are currently available for purchase at the time of writing.

As mentioned before, you may recognize some of the racks listed in Table 5.5, since many of these products include multiple effects.

Type	Manufacturer or publisher	Model	Remarks
Pedal	Boss	RT-20	Four different sounds – Overdrive for guitar or keyboard
	Digitech	Ventura Vibe	Rotary or vibrato effect – 4 settings – 2 inputs – 2 outputs
	Dunlop	Rotovibe	Expression pedal and rotary – Vibrato and chorus mode
	Dunlop	UV-1 Uni-Vibe	Three settings – Chorus/vibrato mode
	Electro Harmonix	Lester G	For guitar – Overdrive emulation – 8 settings – Stereo/mono output
	Fulltone	Deja'Vibe CS-MDV-1	Two settings – Vibrato/chorus and modern/vintage modes
	Hughes Kettner	Rotosphere	12AX7 pre-amp tubes – Stereo/mono
	Leslie	Leslie digital pedal	Four types of speaker and amp (122, 147, 18V, PR40)
	Leslie–Hammond	G Pedal Leslie	Three types of speaker (122, 147, preset)
	Line 6	Roto-machine	Three types of speaker (122, 145, L16) – Configurable transition time
	Neo Instruments	Ventilator II	For organ, guitar and keyboard – Overdrive – 2 inputs – 2 outputs
	Pigtronix	RSS Rototron	Analog – For guitar – 4 settings – 2 inputs – 2 outputs
	Tech 21	RotoChoir	Six settings – Overdrive – Stereo output
	Strymon	Lex	Four settings – Boost function – Stereo output

Rack	Alesis	Nanoverb	Multieffects
	Avid	Eleven Rack	Multieffects – Guitar amp simulators
	Dynacord	CLS 222	Dedicated to the Leslie effect
	Lexicon	MX Series	Digital multieffects
		MPX Series	Digital multieffects
	TC Electronic	M-One	Multieffects
	Zoom Studio	1201	Multieffects
Software	GSI	MrDonald	(freeware) Windows: VST
	Nubi3	Spinner LE	(freeware) Windows: VST
	PSP Audioware	PSP L'otary2	Windows: VST; AAX; RTAS OSX: VST; AAX; RTAS; AU
	MeldaProduction	MVintageRotary	Windows: VST; VST3; AAX OSX: VST; VST3; AAX; AU
	Mda	Leslie	Windows: VST OSX: VST; AU
	BetaBugsAudio	Spinbug	(freeware) Windows: VST
	Xils-Labs	LX122	Windows: VST

Table 5.5. *Models of rotary effects*

5.4.5. *Leslie speakers and sound recording*

One way of recording a Leslie speaker is to use two static wide-band cardioid pickups on either side of the speaker to capture the trebles emitted by the horn.

Importantly, these pickups should not be placed facing each other, but at a slight angle. This avoids any problems due to antiphase phenomena[8] that might eliminate certain frequencies.

To record the rotating drum, which is dedicated to low frequencies, we can place a third microphone in front of the lower vents, angled downwards, carefully ensuring that it cannot cross phases with the top microphones.

---

8 Some DAWs (Digital Audio Workstations) offer integrated plugins (SurroundScope, Voxengo PHA-979, etc.) and tools for managing the phase.

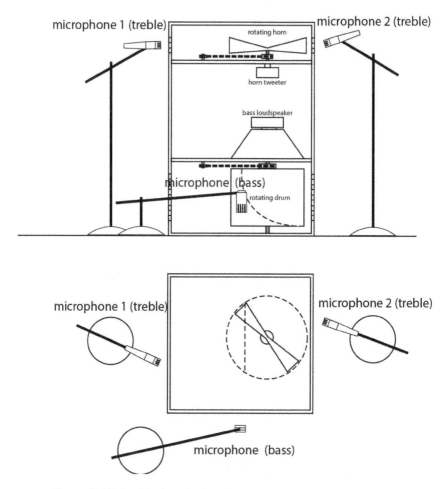

Figure 5.18. *Positioning of microphones to record a Leslie speaker*

Be careful – in "tremolo" mode, the rapid rotation of the drum or horns can generate a hiss that disrupts the microphone recordings. Pop filters can help if you are encountering this issue.

Finally, when the musician stops the rotation of the speakers, there is no reason to expect that the microphones will be in the right place relative to the horns or rotor. An extra dynamic microphone can be placed very close or even inside the speakers (by removing one of the panels, for example the rear panel) to record the inner reflections of the speaker and improve the dynamics.

5.5. *Ring modulation*

This effect is based on a more sophisticated amplitude modulation technique.

Its name comes from the analog circuit used to create the effect, which consists of four diodes arranged in a ring.

Ring modulation was used extensively by electroacoustic (concrete) recording studios in the 1950s–1970s. The composer Karl Stockhausen[9] was passionate about the effect.

5.5.1. *Theoretical principles*

There are several types of analog ring modulation based on circuits with different electronic components. Examples include multiplication-based ring modulation and ring modulation based on "snap-off" diodes.

Figure 5.19 shows the different signal shapes obtained by each method.

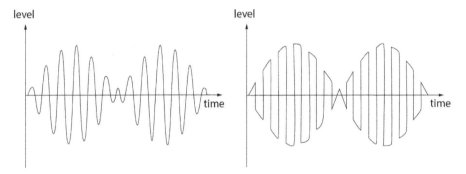

Figure 5.19. *Ring modulation by multiplication (left) and with snap-off diodes (right)*

In practice, in musical applications, the input signal is multiplied together with another signal generated by the modulator. This secondary signal is typically a sine wave. This creates two new signals – one is given by the sum of the input signal and the sine wave, and the other is given by their difference.

---

9 Karl Stockhausen, 1928–2007, German composer who specialized in electroacoustic music and sound spatialization.

For example, if the input signal is a 600-Hz triangle wave and the modulation is a 200-Hz sine wave, the outputs are two signals with frequencies of 400 Hz and 800 Hz (600 – 200 Hz and 600 + 200 Hz), as shown in Figure 5.20.

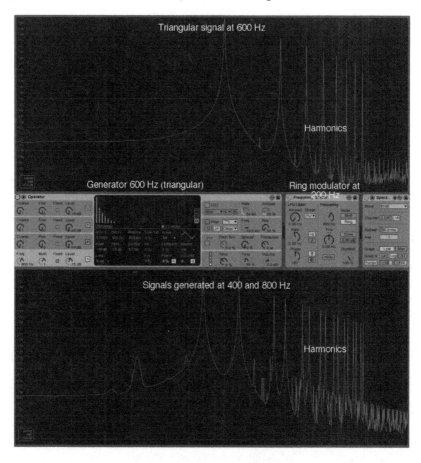

Figure 5.20. *Principle of a ring modulator. The top spectrum analyzer shows a 600-Hz triangle wave. The bottom one shows the output of the ring modulator after combining with a 200-Hz sine wave. The middle panel shows the ring modulator and the spectrum analyzer (in Ableton Live). For a color version of this figure, see www.iste.co.uk/reveillac/soundeffects.zip*

Note that if the modulation frequency is lower than 20 Hz, the modulator is similar to a LFO. In this case, the resulting effect is relatively similar to a tremolo effect.

Most pedals and plugins dedicated to ring modulation are tricky to master. Their settings typically include:

- the choice of modulation waveform;
- the modulation frequency;
- the amplitude of the modulation (*width* or *depth*);
- the frequency of the LFO, if applicable;
- the mix between the original signal and the processed signal (*dry/wet*).

5.5.2. *Examples of ring modulators*

In Table 5.6, you can find a few examples of ring modulators, pedals and plugins.

Type	Brand or publisher	Model	Remarks
Pedal	Way Huge	WHE606 RingWorm	Digital – 5 settings – Input for an expression pedal
	Moog	MF-102 Ring Modulator	Oscillator from 2 Hz to 4 kHz – Waveform: sine and square – Integrated LFO
	Electro Harmonix	Ring Thing	Nine presets – Various settings options
	Moog	MF Ring	Analog – 3 settings
	Digitek	DOD FX13 Gonkulator	1993 – Includes distortion
Plugin	Wok	Ring-O	(freeware) – Windows: VST
	Ndc Plugs	The modulator 2	(freeware) – Windows: VST
	MeldaProduction	MRingModulator	(freeware) – Windows: VST; VST3; AAX OSX: VST; VST3; AAX; AU
	SonicXTC	Ring modulator 101	(freeware) – Windows: VST
	Physical Audio	PA2 Little ringer	(freeware) – OSX

Table 5.6. *Models of ring modulators*

5.6. Final remarks

Where should modulation effects be placed in the audio flow? They are usually placed after distortion, wah-wah, dynamic effects, but before time effects.

Chorus is the most popular and widely used effect, as it is easy to use. Guitarists are very fond of using it to give their sound a rich and thick quality.

Chorus can of course be applied to any type of sound, but it is most effective on signals that are rich in harmonics, which explains why guitarists love it so much.

Phaser and flanger are closely related, although the latter is more precise and subtle. Flanger tends to enlarge the space, whereas phaser tends to fill it. Either way, both effects very quickly create a very specific color when applied overly generously.

Due caution is required – although they make the sound warmer, phaser and flanger tend to sacrifice clarity and impact. You need to find the right compromise between purity, depth of shimmer and dynamics.

Rotary is usually designed for electronic organs, but guitarists were also quick to add it to their repertoire. Although it can be used without moderation with organs, as we have become completely accustomed to how it sounds, some judgment is required with guitar. Rotary does not necessarily suit every style of music.

Finally, ring modulation is a truly unique effect, capable of the best but also the worst. It can create a very synthetic sound rendering, especially with guitar, bass, percussion or piano. There is a good reason it is used by so many synthesizers.

6

Frequency Effects

As their name suggests, frequency effects act upon the frequency of the audio signal, for example by modifying it or varying it. Some of the first effects used by musicians were frequency effects, in the earliest generations of analog and later digital electronics.

The processes and techniques on which these effects are based are often complex. More recent software versions have allowed remarkable advancements to be achieved and highly sophisticated audio processing techniques to be developed.

6.1. Vibrato

The vibrato effect is a cyclic variation of the pitch and hence fundamental frequency of the signal.

Most musicians achieve this effect by directly manipulating the instrument itself, by executing a certain playing technique, usually for very short durations (a few notes or at most a few bars).

It is worth noting that in some cases, especially for vocals and wind instruments, vibrato affects the intensity more than the pitch.

On some electric guitars, vibrato is achieved with a device mounted near the bridge that modifies the tension of the strings to vary the frequency of each note.

Electronic or digital versions of the vibrato effect instead tend to be used over longer periods to create a certain color, style or atmosphere. Vibrato was one of the first effects to be implemented electronically.

6.1.1. *Theoretical principles*

A low-frequency oscillator is used to cyclically vary the frequency of the audio signal up and down, as shown in Figure 6.1.

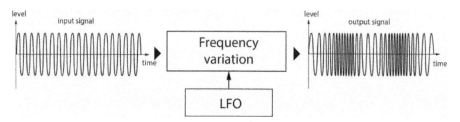

Figure 6.1. *Principle of the vibrato*

The difference between the lowest and highest frequencies determines the depth of the vibrato.

6.1.2. *Settings*

Vibrato can have several user-configurable parameters. Pedals usually only have a few configurable parameters, but software vibratos (plugins) can have many.

Here is a non-exhaustive list of the most common settings:

– the *depth*: determines the intensity of the vibrato;

– the *speed* (or *rate*): determines the frequency of the vibrato;

– the *rise time*: defines the time required for the vibrato to reach the chosen maximum intensity (depth);

– the *tone*: shifts the effect toward lower or higher frequencies;

– the *shape* of the variation signal: defines the waveform of the vibrato, usually ranging from a sine wave to a triangle wave.

Frequency Effects 133

Figure 6.2. *The "MVibrato" plugin by MeldaProduction*

6.1.3. *Examples of vibrato*

Figure 6.3. *Three models of vibrato pedal: "Viper" by T-Rex, "VB-2w" by Boss and "NuVibe" by Korg*

Some manufacturers combine the vibrato and tremolo effects into a single pedal.

REMARK.– People often fail to properly distinguish between the vibrato and tremolo effects. This confusion has been perpetuated by some pedal manufacturers, who frequently misuse the two terms. Many instrument amps also list vibrato on their controls for effects that are in fact tremolos.

Type	Manufacturer or publisher	Model	Remarks
Pedal	Boss	VB-2W Vibrato Waza Craft	Analog – two types of vibrato – Connection for an expression pedal
	Digitech	Ventura Vibes	Vibrato and rotary
	Electro Harmonix	Tube Wiggler	Vibrato and tremolo – four types of effect – 12AX7WB analog with tubes
	Electro Harmonix	The Worm	Four effects: phaser, tremolo, vibrato and modulated wah
	Electro Harmonix	Good Vibes	Connection for an expression pedal
	Fulltone	DejaVibe 2	Vibrato and chorus – two modes: modern and vintage
	Korg	NuVibe	Chorus and vibrato – Reissue of the famous "Uni-vibe" – Allows the waveform to be configured
	Malekko	Vibrato	Analog
	TC Electronic	Shaker Vibrato	Two types of vibrato – USB port
	T-Rex Engineering	Viper	Vibrato and rotary
Rack	Avid	Eleven Rack	Multieffects – Guitar amp simulators
	Digitech	GSP1101	Multieffects – Guitar amp simulators
	Lexicon	MX Series MPX Series	Digital multieffects Digital multieffects
	Line 6	Pod Pro	Multieffects – Guitar amp simulators
Software	Audio Damage	PulseModulator	(freeware) – Windows: VST OSX: VST; AU
	MeldaProduction	MVibrato	(free) – Windows: VST; VST3; AAX OSX: VST; VST3; AAX; AU
	Plug & Mix	Vibrator	Windows: VST; AAX OSX: VST; AAX; AU
	Two Circuits	Audiovibrato	(freeware) – Windows: VST

Table 6.1. *Examples of vibrato*

6.2. Transposers

These effects modify the tone of the sound. There are several variations, including:

– *octaver;*

– *pitch shifter;*

– *harmonizer;*

– *autotune.*

Although these effects can be applied to any instrument, the first three are most commonly used by guitar players. They were invented with the goal of enriching the harmonics of a sound passage, similar to musical arrangements that overlay the same musical phrase at different pitches (octave, fifth, third, etc.).

REMARK.– These effects are also often confused, and consequently different models of transposer work differently. The pitch shifter effect can be viewed as a harmonizer, and so can the octaver. Digital technology has muddied the waters even further, providing an even more extensive range of sound processing options.

6.2.1. *Octaver*

This effect divides the frequency of the input signal by two. Musically, this lowers the sound by one octave. By applying the same operation twice, we can reduce by two octaves.

Figure 6.4. *The pedal "Octavia" by Tycobrahe, a commercial version of the pedal used by Jimi Hendrix, marketed in the 1970s*

One of the first ever octavers was the famous "Octavia" by Tycobrahe, designed in 1967 by Roger Mayer, Jimi Hendrix's sound technician. Octavia not only added an octave to the original sound, but also applied a saturation effect (*fuzz*). The sound produced by this fully analog pedal was far from clear or precise, but its unique effect was very characteristic.

6.2.1.1. *Examples of octavers*

Figure 6.5. *Three models of octaver pedal: "Whammy" by Digitech, "Super Octaver OC-3" by Boss and "Octavio JH-OC1" by Dunlop*

Type	Manufacturer or publisher	Model	Remarks
Pedal	Boss	OC-3 Super Octave	Three modes including a polyphonic effect
	Dane Electro	French Toast Octave Distortion	Distortion and octaver
	Digitech	Whammy	Nine harmonizer presets – six whammy presets – two detune presets – MIDI input
	Dunlop	JH-OC1 – Octavio	Two settings: level and fuzz
	Dunlop	JH3S	Fuzz and octaver – two settings: volume and tone
	Electro Harmonix	Octavix	Octaver and fuzz – three settings: volume, octave, and boost
	Electro Harmonix	Octave Multiplexer	Three settings: high filter, blend, and bass filter – Switch sub on/off
	Eventide	PitchFactor	Multi-effects: 10 different effects including octave
	Foxx	Tone Machine	1972 – Octaver and fuzz – Analog – one switch: fuzz/octafuzz – three settings: volume, sustain, fuzz

	Fulltone	Octafuzz OF-2	Copy of the famous "Octavia" by Tycobrahe – two settings: volume and boost
	Fulltone	Ultimate Octave	Fuzz and octaver – Switch: octave-up – three settings: volume, tone, and fuzz
	Joyo	JF-12 Voodoo Octave	Octaver and fuzz – three settings: fuzz, tone, and volume – two switches: fuzz and octave
	Movall	Octopuzz	Two settings: level and tone
	MXR	Custom Shop Machine Fuzz	Fuzz and octaver – Analog – four settings: volume, tone, fuzz, sub
	Roger Mayer	Octavia Classic	Reissue from the 1960s
	TC Electronic	Sub 'N' Up Octaver	PitchShifter, Octaver, Harmonizer – USB port
	Tycobrahe	Octavia	1970 – Analog – The gold standard – Hard to find
	Voodoo Lab	Proctavia	Fuzz and octaver – two settings: volume and boost
Rack	Avid	Eleven Rack	Multieffects – Guitar amp simulators
	Digitech	GSP1101	Multieffects – Guitar amp simulators
	Lexicon	MX Series	Digital multieffects
		MPX Series	Digital multieffects
	Line 6	Pod Pro	Multieffects – Guitar amp simulators
Software	Antares	Harmony Engine EVO	Pitch shifter and time stretcher Windows: VST; AAX; RTAS OSX: VST; AAX; RTAS; AU MIDI features
	Audiowish	Octave Shifter 2	(freeware) OSX: AU
	Chris Hooker	Octaver OC-D2	(freeware) Windows: VST
	TbT	Octaver 12B	(freeware) Windows: VST
	Waves	UltraPitch	Windows: VST; VST3; AAX; RTAS OSX: VST; VST3; AAX; RTAS; AU

Table 6.2. *Examples of octavers*

6.2.2. *Pitch shifter*

This effect modifies the pitch of a sound signal while keeping its duration constant. By convention (and in most cases), pitch shifters change the pitch by a constant number of semitones over one or more octaves.

Analog pitch shifters do not work well with polyphonic sound sources. This effect is primarily designed for mono signals.

This is no longer a problem with digital models. Virtually all digital models support polyphonic sound.

Figure 6.6. *Three pitch-shifter pedals: "Pitch Fork" by Electro Harmonix, "PS-5" by Boss and "Particle" by Red Panda*

Type	Manufacturer or publisher	Model	Remarks
Pedal	Boss	PS-2	1987–1994 – Pitch shifter and delay (2 s max) – six different modes
	Boss	PS-3	1994–1999 – Pitch shifter (–2 oct to +2 oct) and delay (2 s max) – 11 different modes
	Boss	PS-5 Super Shifter	1999–2011– Digital – Polyphonic – Thirds, fourths, fifths, sevenths, one octave, two octaves
	Eventide	PitchFactor	Pitch shifter and delay (1.5 s max) – four diatonic pitch-shifting tracks – 10 different modes – 100 presets
	Electro Harmonix	Pitch Fork	2014 – Digital – Polyphonic – ±3 octaves – three modes
	Red Panda	Particle	Pitch shifter and granular delay (900 ms max) – eight different modes
Rack	Boss	RPS-10	Two settings: level and fuzz
	Digitech	Intelligent Pitch-Shifter	1987 – Pitch shifter (–1 oct to +1 oct) and delay (800 ms max)

Software	Eventide	H949	1977 – Pitch shifter, harmonizer, and deglitching
	Aegan Music	Pitchproof	Harmonizer and pitch shifter Windows: VST
	Aegan Music	Pitchometry	Windows: VST OSX: VST; AU
	AuraPlug	Whamdrive	(freeware) – Pitch shifter and distortion 14 different modes Windows: VST OSX: VST; AU
	kiloHearts	Pitch shifter	Windows – OSX: VST
	Morfiki	grANALiser	(freeware) – Windows: VST
	SoundToys	Little Alterboy	Pitch and formant shifting Windows: VST; AAX OSX: VST; AAX; AU
	Waves	SoundShifter	Windows: VST; VST3; AAX; RTAS OSX: VST; VST3; AAX; RTAS; AU

Table 6.3. *Examples of pitch shifters*

6.2.3. Harmonizer

Harmonizers are similar to another effect that we will discuss in Chapter 8, the *delay* effect. The input signal is routed to both the output and a delay line that changes its frequency. This part of the processed signal is then mixed with the original input signal, while ensuring that both signals are synchronized.

Harmonizers can detune the original signal by shifting it to a higher or lower frequency over one or several octaves.

Figure 6.7. *The H910 harmonizer by Eventide (1975–1984)*

Unlike pitch shifters, harmonizers take into account the key in which the notes are being played by adjusting the number of semitones in the shift to be consistent with the key and the choice of interval (thirds, fifths, octaves, etc.).

For example, in C major, if the harmonizer is set to thirds, it will apply the following shifts, as presented in Table 6.4.

Original note	Note generated by the harmonizer	No. of semitones	Original note	Note generated by the harmonizer	No. of semitones
C	E	4	F#	A	3
C#	E	3	G	B	4
D	F	3	G#	B	3
D#	F#	3	A	C	3
E	G	3	A#	C#	3
F	A	4	B	D	3

Table 6.4. *The notes generated by the harmonizer to create a third in the same key. The number of semitones depends on the note*

REMARK.– Some pitch shifters have harmonizer functions and vice versa.

Figure 6.8. *Three harmonizer pedals: "AHAR-3" by Tom's Line Engineering, "Harmonist PS-6" by Boss and "Harmony" by Hotone Audio*

Type	Manufacturer or publisher	Model	Remarks
Pedal	Digitech	HarmonyMan	Harmonizer – Whammy octaver (pitch shift)
	DOD	Meatbox 2015	Harmonizer – Whammy octaver (pitch shift)
	Eventide	H9 Harmonizer	Multi-effects: ModFactor, Pitchfactor, TimeFactor – USB port – iOS app
	Hotone Audio	Harmony	Harmonizer – Whammy octaver (pitch shift) – 24 semitones
	Morpheus	Bomber	Harmonizer – Octaver – Whammy (pitch shift)
	Tom's Line Engineering	AHAR-3 Harmonizer	Harmonizer and pitch shifter
Rack	AMS Neve	DMX 15-80S	1978 – Delay and digital harmonizer
	Eventide	H910	1975–1984 – First digital harmonizer – four delays – CV input
	Eventide	H3000 Ultra Harmonizer	Multi-effects – 21 algorithms – 600 presets
	Eventide	H7600 Ultra Harmonizer	Multi-effects – 1,000 algorithms – 230 effects blocks – Windows and OSX interface
Software	Aegan Music	Pitchproof	Harmonizer and pitch shifter – Windows: VST
	Eventide	H3000	Digital emulation of the H3000 Windows: VST; AAX; RTAS OSX: VST; AAX; RTAS; AU
	Eventide	H910	Digital emulation of the famous H910 Windows: VST; AAX; RTAS OSX: VST; AAX; RTAS; AU
	Martin Eastwood	Duet	(freeware) – Windows: VST
	Waves	UltraPitch	Windows: VST; VST3; AAX; RTAS OSX: VST; VST3; AAX; RTAS; AU
	Waves	Doubler	Pitch detune, chorus, delay Windows: VST; VST3; AAX; RTAS OSX: VST; VST3; AAX; RTAS; AU

Table 6.5. *Examples of harmonizers*

6.2.4. Auto-Tune

Auto-Tune was originally designed to correct out-of-tune vocals. It was created in 1997 and has since also become popular as a fully fledged vocal synthesis instrument, inspiring the development of new tools that can be classified as voice processors.

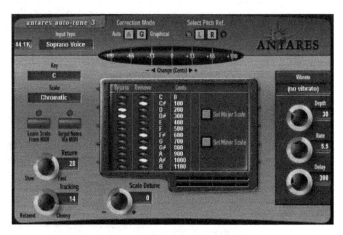

Figure 6.9. Antares Auto-Tune Version 3 for DirectX, Windows. For a color version of this figure, see www.iste.co.uk/reveillac/soundeffects.zip

6.2.4.1. History

Auto-Tune was created by Harold (Andy) Hildebrand, whose field of expertise amusingly had nothing to do with music. He worked for the petroleum industry, and specialized in determining the feasibility and viability of exploiting oil deposits.

As part of his work, he developed a method based on autocorrelation[1] that allows acoustic waves to be transmitted underground. After perfecting his method, he sold it to the oil company Exxon, and retired.

Some time later, at a dinner party – as he tells the story – somebody challenged him to develop an application that makes you always sing in tune. Using exactly the same autocorrelation technique that he had used for his work, he developed the software program "Auto-Tune" in 1996.

---

[1] Autocorrelation is a mathematical tool used for signal processing that allows patterns in a signal to be detected, among other things.

Precursory tools had long been available, such as the "Voder", invented in 1937 by the engineer Homer Dudley[2], which modified the voice of singers, and the "Sonovox", which was created in 1940. The Sonovox can be seen in the movie "You'll Find Out" starring Kay Kyser.

Figure 6.10. *The "Voder" invented by Homer Dudley*

But it was the German physicist Werner Meyer-Eppler, founder of the Cologne Studio for Electronic Music, who was truly responsible for propelling voice effects onto the musical scene in the late 1950s.

In 1968, because of technological advancements, the famous synthesizer manufacturer Moog released their first "vocoder" for sale. This effect can be heard in many musical tracks from this period, including "Autobahn" by Kraftwerk.

---

2 Homer Dudley, 1896–1987, American engineer, creator of the first speech synthesizer at Bell Labs, which was used to encrypt voice communications during World War II. He also invented the "Vocoder" in 1939.

Figure 6.11. *Sixteen-channel vocoder by Moog*

Even though they use similar technology, the "vocoder" and "Auto-Tune" are not the same. The former generates sounds using a keyboard to create vocal harmonies, and the latter processes vocals using dedicated software.

The first commercial version of Auto-Tune was developed in 1997 by Audio Antares Technologies, a company founded by H. Hildebrand.

One of the first hits to use the effect was "Believe", performed by the singer Cher in 1998. Although Auto-Tune was originally intended as a tool for singing in tune, this song used it as a musical effect in its own right. Many other tracks would later follow suit.

Today, Auto-Tune is no longer just a tool for cleaning up vocals, but has become an entirely new form of instrument.

Figure 6.12. *Version 8 of the Auto-Tune plugin by Antares. For a color version of this figure, see www.iste.co.uk/reveillac/soundeffects.zip*

6.2.4.2. Autotune effects and voice processors

Table 6.6 contains a list combining autotune effects, vocoders and voice processors. The "other" category consists of "desktop" processors that can be placed on a table or attached to a microphone stand. They are usually specifically designed for vocals.

Figure 6.13. *Four voice processors: "Perform-V" by TC-Helicon, "SVC-350" by Roland, "TA-1VP" by Tascam and "VE-2" by Boss*

Type	Manufacturer or publisher	Model	Remarks
Pedal	Boss	VE-2	Voice processor – three presets – 24 types of harmonization
	Boss	VE-8	Voice processor for singers accompanied by acoustic guitar
	Electro Harmonix	Voice Box	Harmonizer and vocoder – nine modes
	Mooer	VEM Box	Voice processor – 54 types of effect
	TC-Helicon	Harmony singer	Voice processor for voice + guitar – USB – Detection of guitar key
	TC-Helicon	VoiceLive3 Extreme	Voice processor – Various modes for voice and guitar
	TC-Helicon	VoiceTone Correct XT	Voice processor – Automatic pitch correction, de-esser, equalizer
Rack	Roland	VP-70	Voice processor and harmonizer
	Roland	SVC-350 Vocoder	Voice processor and vocoder – Inputs for microphone, guitar, synthesizer
	Tascam	TA-1VP	Voice processor – Includes Antares Auto-Tune – Microphone modeling – MIDI
	TC-Helicon	VoiceWorks	Voice processor – Harmonizer – Various features – MIDI

Software	Antares	Auto-Tune Live	The live version of the Antares Auto-Tune effect Windows: VST; AAX OSX: VST; AAX; AU
	Antares	Auto-Tune 8	The latest version[3] of Antares Auto-Tune Windows: VST; AAX OSX: VST; AAX; AU MIDI features
	Antares	Auto-Tune EFX 3	Simplified version of Auto-Tune – Limited corrections and settings Windows: VST; AAX OSX: VST; AAX; AU
	g200kg	Kerovee 1.61	(freeware) – Windows: VST
	GVST	Gsnap	Pitch corrector (freeware) – Windows: VST
	Oli Larkin	Autotalent	(freeware) – Windows: VST
	SoundToys	Little AlterBoy	Hard-tune – Pitch shifter – Harmonizer – Various other effects Windows: VST; AAX OSX: VST; AAX; AU
	Waves	SoundShifter	Voice processor Windows: VST; AAX OSX: VST; AAX; AU
Other	Roland	VT-3	Voice processor Desktop or microphone stand controller
	TC-Helicon	Perform-V	Voice processor Mounted on a microphone stand
	TC-Helicon	VoiceLive Touch 2	Voice processor Desktop or microphone stand controller

Table 6.6. *Examples of voice processors, vocoders and autotuned effects*

6.2.4.3. *Why do we need autotune?*

You might be wondering: what is the point of using autotune?

---

3 Version 8 is the latest version at the time of writing this chapter (February 2017).

Autotune has been widely criticized by the media, but it is used by 90% of all professional recording studios. Here are four arguments in favor of autotune that I personally find convincing:

– when used in moderation, autotune unquestionably improves the quality of vocals without distorting them;

– voice modification represents a simple way of enhancing or deliberately distorting vocals for stylistic effect, but does not just make anybody into a good singer;

– autotune saves time when processing renderings of vocals;

– it can be used as a replacement for vocoder, allowing vocals to be "robotized", giving them a strongly artificial quality.

Opinions vary, but auto-tune is here to stay. Like many new technologies, progress is difficult to stop, no matter which side you are on. Try it out, and decide for yourself!

6.2.4.4. How to use Antares Auto-Tune

The next few sections give a brief introduction to Antares Auto-Tune and some of its parameters.

After installing Antares Auto-Tune in your favorite DAW (Digital Audio Workstation), load the voice segment that you want to process onto a new track. You can also open a multitrack project with one or several dedicated voice tracks.

Figure 6.14. *The opening window of the Antares Auto-Tune 8 plugin. For a color version of this figure, see www.iste.co.uk/reveillac/soundeffects.zip*

On this track, load the Antares Auto-Tune effect, which might be in a VST, AU or AAX format.

A plugin window will open, displaying various settings, as shown in Figure 6.14.

Before processing our voice track, let us take a moment to explore this interface, starting with the top ribbon.

The drop-down list INPUT TYPE allows you to select the register that best matches the voice track. SOPRANO is the default option.

– SOPRANO: soprano voice;

– ALTO/TENOR: alto or tenor voice;

– LOW MALE: bass voice;

– INSTRUMENT: for instrumental tracks (Auto-Tune also works on instruments, but this is less popular), such as a guitar;

– BASS INSTRUMENT: for instruments with lower registers, for example a bass guitar.

Next, the TRACKING control ranges from RELAXED to CHOOSY, allowing us to specify whether the voice contours are clean and isolated (recording studio), or whether there is ambient background noise (live recording).

For studio environments, we will tend to use RELAXED, whereas in live environments we will tend to prefer CHOOSY, since we will need to be more selective. A good default for conventional sound recording is simply 50%.

Figure 6.15. *Top ribbon of Auto-Tune 8*

The PITCH REF switch specifies which channel should be used to differentiate between the notes of the melody for stereo signals. If the voice is equally balanced across both channels, we do not need to specify a preference for left (L) or right (R).

The LOW LATENCY button activates a mode that reduces the latency time to unnoticeable levels (for live applications), at the cost of the processing quality. Do not use this option for postproduction (studio processing).

The KEY, SCALE and SCALE DETUNE block are used to define the key and scale of the voice track.

KEY specifies the key, from C to B, and SCALE allows the scale or mode to be selected, either CHROMATIC, MAJOR or MINOR. The SCALE DETUNE button is used to define the primary chord with fundamental note A. By default, this is 440 Hz, but it can be adjusted by –1 to +1 semitone (100% = 1 semitone).

The next block has the TRANSPOSE, THROAT LENGTH and FORMANT buttons:

– the TRANSPOSE function modifies the overall tuning (pitch) of the melody over an interval of ± 1 octave;

– the THROAT LENGTH function controls the timbre of the vocals, e.g. switching between a child's voice to a woman's or a man's. This function only works when FORMANT is turned off;

– the FORMANT button attempts to preserve the characteristics of the formant[4] of the singer's voice, independently of all other settings.

The final block consists of the CORRECTION MODE switch and the OPTIONS button:

– the CORRECTION MODE function switches between automatic voice correction mode (AUTO) and graphical editing mode (GRAPH);

– the OPTIONS button opens a dialog window with general Auto-Tune settings: buffer size, number of undos, button mode, etc.

Figure 6.16. *PITCH CORRECTION CONTROL settings*

---

4 The spectral envelope that characterizes an acoustic signal like a person's voice along their vocal tract, including resonance. The formant is specific to each individual.

Next, let us look at the subblock for PITCH CORRECTION CONTROL, which controls the overall pitch of the voice using five controls:

– RETUNE SPEED: defines the reactivity of the pitch correction in milliseconds. The lower this value, the more natural the voice will seem, and vice versa;

– CORRECTION STYLE: restricts Auto-Tune's corrections to allow the vocalist to retain as much expression as possible (FLEX-TUNE). The NONE position means that no corrections are made, and the CLASSIC position allows Auto-Tune to operate normally;

– HUMANIZE: this function differentiates between short notes and long or sustained notes by slowing down the *retune speed* on longer notes, which "humanizes" the voice, making it seem more natural;

– NATURAL VIBRATO: defines the degree to which Auto-Tune corrects notes in the vocals with vibrato (notes that oscillate around the correct pitch);

– TARGETING IGNORE VIBRATO: instructs Auto-Tune to disregard vibrato in the vocals.

REMARK.– When TARGETING IGNORE VIBRATO is turned off, Auto-Tune tends to make the melody sound out of tune if the vocals include strong vibrato effects.

Next, we will examine the central group, which by default contains buttons corresponding to each of the 12 notes of the chromatic scale, from C to B, and other buttons for configuring the scale or musical mode.

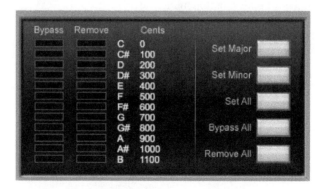

Figure 6.17. *Group for selecting notes and modes*

By clicking on the buttons in the REMOVE column lined up with each note, you can remove them from pitch correction in the chosen key.

By clicking on one of the BYPASS buttons, you remove pitch correction from that note regardless of the chosen key. Auto-Tune acts as if these notes do not exist:

– the SET MAJOR button sets the scale to major mode;

– the SET MINOR button sets the scale to minor mode;

– the SET ALL button adds every note in the chosen scale to pitch correction;

– the BYPASS ALL button excludes every note from pitch correction;

– the REMOVE ALL button removes every note from the chosen scale from pitch correction.

Now that we have been introduced to some of the commands, we can process our first audio sequence:

– first, select the primary register of the voice in INPUT TYPE. This is usually SOPRANO for female voices or children and ALTO/TENOR for male voices;

– next, set the TRACKING to 50 (middle position) and make sure that LOW LATENCY is switched on (blue);

– choose the key and scale of the melody, e.g. C major, G minor, etc. from the KEY and SCALE drop-down lists. If you do not know the key or scale of your track, simply choose C and CHROMATIC;

– if necessary, adjust the SCALE DETUNE overall tuning. In the United States, musicians usually tune A to 440 Hz, with some exceptions. For example, you can sometimes find instruments and orchestras with A tuned to 442 Hz;

– start playing the track with Auto-Tune, and listen carefully so that you can configure the reactivity (RETUNE SPEED). You need to make sure that as many notes as possible are captured by the correction space without causing the voice to seem unnatural;

– adjust the HUMANIZE setting to fine-tune the quality of sustained notes;

– you can also adjust the inflections of the vocals to improve their spontaneity with the CORRECTION STYLE parameter, although simply setting it approximately to FLEX-TUNE will usually be your best bet;

– if necessary, you can polish the final rendering by adjusting any vibrato with the NATURAL VIBRATO button.

REMARK.– If you chose CHROMATIC because you are not sure which key the vocals are in, you can deselect specific notes in the middle panel in the REMOVE column to prevent them from being processed.

Auto-Tune works by always correcting to the closest note in the specified scale.

For example, if the melody is written as A, but the singer is very out of tune and sings closer to G, Auto-Tune will replace this A by G if G is in the scale (center panel). To avoid this problem, you can remove G from the scale.

REMARK.– You usually will not be able to correct an entire voice track in a single operation. I recommend working on short sequences of a few notes or a few dozen notes at most. This will allow you to make corrections that are imperceptible in the final mix. Auto-Tune is a powerful tool, but requires plenty of patience and attention to achieve good vocal renderings.

Auto-Tune also has a graphical voice editing mode that can be accessed by setting CORRECTION MODE to GRAPH. The graphical editor enables you to correct each phoneme or syllable in the voice melody. This approach is used by professionals to achieve pitch-perfect vocals that sound as natural as possible. The vocals are often split into several tracks in the DAW and recombined later. Auto-Tune can work separately with both channels of a stereo signal.

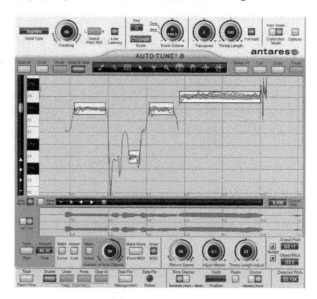

Figure 6.18. *Auto-Tune's graphical editing mode for making extremely fine adjustments to vocals. For a color version of this figure, see www.iste.co.uk/reveillac/soundeffects.zip*

The full functionality of the graphical editor goes well beyond the scope of this book. You can find out more from the web links and bibliography at the end of the book.

6.2.4.5. *Synthetic voices with Auto-Tune*

Today, Auto-Tune has become a very fashionable way of deforming vocals to give them a synthetic feel. As we mentioned earlier, Cher was the first singer to do this, in her song "Believe". Since then, many other artists have taken advantage of this technique, most notably rappers.

To recreate this effect, repeat the steps as in section 6.2.4.4, but set SCALE DETUNE to a value between 445 and 460, RETUNE SPEED to FAST and optionally set PITCH AMOUNT in the CREATE VIBRATO block to a value between 20 and 30.

REMARK.– These values are only guidelines. You will need to fine-tune the settings to achieve fluid results with a high-quality artificial and synthetic rendering.

6.2.4.6. *For fun: autotune effects on smartphones and tablets*

Although they really are not much more than gadgets, Auto-Tune's popularity inspired the development of several fun little apps:

– MicDroid for Android: autotune effect (freeware);

– Songify by Smule for Android and iOS: turns your lyrics into songs (commercial);

– Auto-Tune Mobile for iOS: Antares Auto-Tune for iPad and iPhone (commercial);

Figure 6.19. *Apps: Auto-Tune by Antares (left) and Voice Synth by Qneo (right) for iOS. For a color version of this figure, see www.iste.co.uk/reveillac/soundeffects.zip*

– I'm T-Pain by Smule for iOS: autotune to sing like T-Pain (commercial);

– Voloco by Resonant Cavity LLC: Auto Voice Tune: harmonizer and karaoke (freeware);

– Voice Synth Free by Qneo: autotune and vocoder (freeware);

– Voice Synth Free by Qneo: more advanced autotune and vocoder (commercial);

6.3. Conclusion

Frequency effects are always capable of achieving a fantastic rendering, but they need to be used in some moderation to avoid stifling the music.

Excessive usage quickly leads to chaos.

However, pitch correction effects have more recently become extremely popular for vocals. Professionals and consumers alike have become accustomed to hearing them everywhere. Personally, I think that it is a bit of a shame. The proliferation of autotune has created a tendency to standardize sounds and depersonalize the individual performance of vocalists.

The "robot voice" autotune trend has been everywhere these past few years. It should hopefully die off soon, given widespread criticism by music critics, the media and professional musicians, all of whom have had enough. Too much is too much!

7

Dynamic Effects

This family of effects acts on the relative level of maximum loudness of the sound signal. Dynamics are described by the dynamic range, which is a dimensionless quantity expressed in decibels. The dynamic range characterizes the level of sound as perceived by an observer, and can be influenced by variations in the acoustics and timbre, which often characterize our perception of the power of the sound. Physically, dynamic effects nuance the sound by modifying the spectral composition of the signal.

Each musical instrument has its own dynamic range, defined by the peak sound level that this instrument is able to generate.

In the context of a recording medium, things work slightly differently. The dynamic range is instead defined by the difference between the zero-signal level and the maximum (peak) level supported by the medium. Above the peak level of the medium, the signal becomes saturated due to the limitations of the medium itself, unrelated to the properties of the original signal.

REMARK.– In a recording medium, the zero-signal level is the minimum level of background noise. For example, with magnetic tape, the zero-signal level is the background noise of the tape itself (always present). Any signal that is weaker than this background noise is physically impossible to record.

Dynamic effects take many forms, e.g. pedals, racks and software components, either directly integrated into a Digital Audio Workstation (DAW) or available as specialized plugins.

7.1. Compression

Compression is widely used by professional recording and sound technicians. The idea is simple. Compression automatically adjusts the level of the sound signal to certain predefined values in order to manage variations in volume.

Compression changes both the quality and our perception of the signal. This is mostly because it strongly affects the *attack* and *release*.

Compression is used at virtually every stage of musical production, whether recording, mixing, reproduction or publishing.

7.1.1. *History*

Before compressors were invented, sound engineers and technicians needed to manually vary the volume to compensate for the fluctuations in the incoming sound signal.

The first compressors were not introduced until the 1950s. They were initially developed for radio. Limiters had been invented earlier, in the 1930s, shortly after vacuum tubes were discovered.

But the widespread adoption of compressors by recording studios would have to wait until the 1960s.

7.1.2. *Parameters of compression*

Usually, compression is only triggered above a certain sound level, called the *threshold*, and expressed in decibels. Below this threshold, the compressor is inactive, leaving the sound completely intact.

Figure 7.1. *Drawmer 1973 studio compressor*

When the threshold is exceeded, dynamic range compression is activated. Its action is governed by the following parameters:

– the input and output levels (gains);

– the compression ratio;

– the attack;

– the release;

– the *make-up gain*.

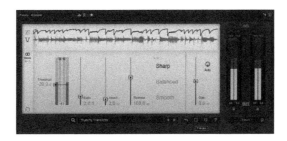

Figure 7.2. *The software compressor "Ozone 7 Vintage Compressor" by Izotrope. For a color version of this figure, see www.iste.co.uk/reveillac/soundeffects.zip*

7.1.2.1. *The compression ratio*

This parameter describes the proportion by which the sound will be compressed. The higher the compression ratio, the greater the compression.

Suppose that n denotes the number of decibels above the threshold. Whenever n increases by x dB, the compressed signal only increases by y dB (beginning at the threshold).

x:y is the compression ratio.

The following equation describes the output level of the compressor:

$$n_o = \frac{y(n_i - s)}{x} + s$$

where:

– n_o: output level in dB;

– n_i: input level in dB;

– s: threshold in dB;

–x: numerator of compression ratio;

–y: denominator of compression ratio.

Let us consider an example.

Suppose that the threshold is 15 dB and the ratio is 2:1. If the input signal is 17 dB, we will have:

$$n_o = \frac{1(17-15)}{2} + 15$$

$$n_o = 16$$

The compressor will reduce the output signal to 16 dB.

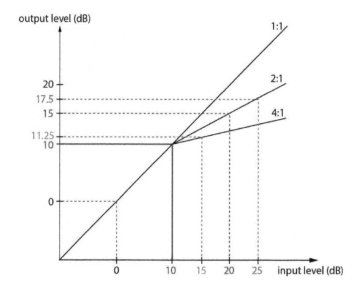

Figure 7.3. *Graph of the output level as a function of the ratio (x:y), the threshold (s) and the input level (n_i). Given a ratio of 2:1, a threshold of 10 dB and an n_i of 20 dB, the output level is 15 dB. For a color version of this figure, see www.iste.co.uk/reveillac/soundeffects.zip*

7.1.2.2. *The attack*

Measured in milliseconds, the attack is the time that elapses before the compressor begins to act. Varying the attack can significantly change the timbre of the signal.

If the attack is too long, the compressor will struggle to stabilize, which negatively affects the signal quality. The attack duration needs to be chosen in such a way as to retain and preserve the *transients*, i.e. the earliest information in the sound wave, namely the initial amplitude of the wave.

Some sounds, like those made by percussion instruments, have extremely obvious and pronounced transients. Other instruments, such as strings, create much softer transients.

If the attack is too short, the signal will be flattened and the percussive components of the signal will be lost.

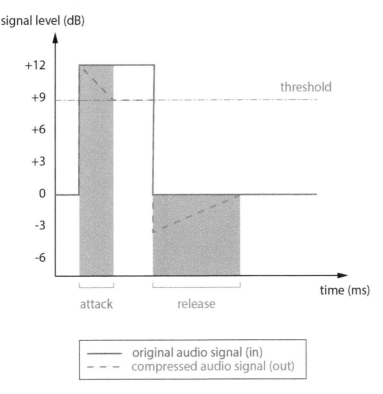

Figure 7.4. *Attack and release during compression. For a color version of this figure, see www.iste.co.uk/reveillac/soundeffects.zip*

7.1.2.3. *The release*

The release is also measured in milliseconds, like the attack. It represents the time that elapses between the moment when the sound passes back below the threshold and the moment when the compressor turns off. The release also strongly influences the timbre of the sound.

Many compressors have an automatic release mode. In this mode, the compressor intelligently adapts to the type of input signal.

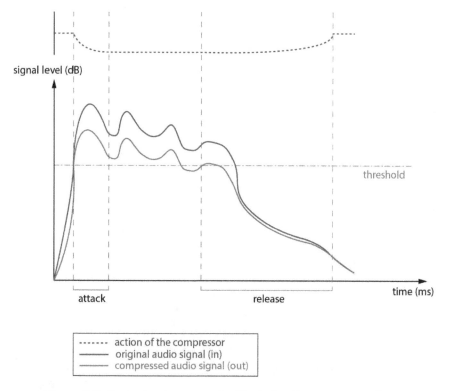

Figure 7.5. *Example of compression illustrating its parameters. For a color version of this figure, see www.iste.co.uk/reveillac/soundeffects.zip*

7.1.2.4. *The make-up gain*

This parameter specifies how to compensate any loss in the signal gain caused by compression. It therefore determines the final gain.

7.1.2.5. *Other functions*

The *knee*: this determines whether the compressor should turn on gradually, which is known as *soft knee*, or abruptly, which is known as *hard knee*;

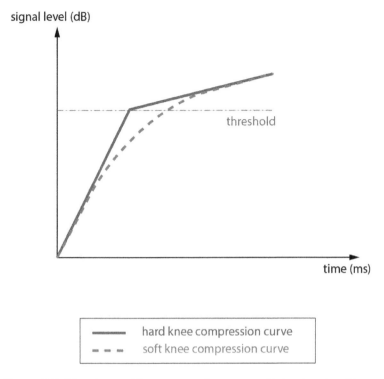

Figure 7.6. *The shape of the compression curve with hard and soft knee*

– the *link* (or *slave*): this function is useful when using groups of multiple compressors (often in pairs), e.g. for a stereo signal. Each group can be configured independently with different settings, or linked together so that every compressor in the group uses the same parameters value. The slave compressor inherits its configuration from the master compressor. Link mode can help to preserve the original character of stereo signals;

– the limiter: this traps the signal below the threshold. Compressors can effectively be viewed as limiters whenever their compression ratio is 10:1 or higher. Limiters are often used in specific situations to ensure that critical thresholds are not exceeded during sound recording, or to prevent damage to the speakers in live setups, and so on;

– the *sidechain* (or *key*): this is a separate device connected to the compressor that uses an external signal to control when the compressor should be activated. Now that software compressors (plugins) have become more widely available, this feature has become easier to find; previously, it was only found in high-end compressors. Sidechains can also include other (internal) processing operations, such as low-pass filters to eliminate *pumping*[1], switching to an insert such as an equalizer (see section 4.2) or a de-esser (see below, section 7.4). One of its very first applications was to lower the volume of music when an announcement is made on the microphone, as is still common practice today;

– the mix: this determines the ratio of original audio input signal to compressed signal in the output of the compressor;

– the *low-cut comp*: this function, included in some compressors, eliminates low frequencies in the sidechain control signal;

– *rotation point compressor*: this function, when available, creates a rotation at the point where the audio signal and the threshold intersect. This amplifies the signal below the threshold and compresses the signal above it.

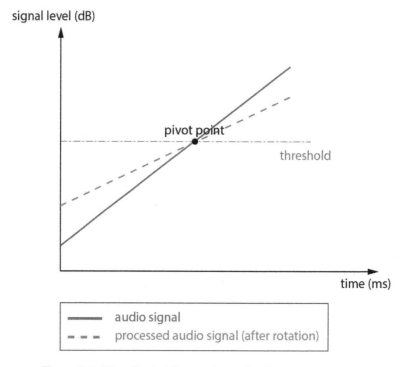

Figure 7.7. *The effect of the rotation point. For a color version of this figure, see www.iste.co.uk/reveillac/soundeffects.zip*

---

1 If the compressor releases too quickly, you can sometimes hear the increase in the sound level after the compressor stops cutting the peaks. This phenomenon is called "pumping".

7.1.3. *Examples of compressors*

Table 7.1 gives a non-exhaustive list of a few compressors, available as pedals, racks or plugins.

Type	Manufacturer or publisher	Model	Remarks
Pedal	Biyang	CO-10 Compress X	Compressor/sustainer for guitar
	BOSS	CP-1X	Multiband compressor for guitar
	Carl Martin	Classic opto compressor	Optical compressor* for guitar
	Diamond Pedals	Compressor	Optical compressor*
	Fender	Micro-compressor	Compressor for guitar
	Marshall	ED-1 The compressor	Compressor for guitar
	MXR	M132 Super comp	Compressor for guitar
	MXR	M76 Studio	Compressor for guitar
	Providence	Velvet VLC-1	Compressor/sustainer for guitar
	Rocktron	Reaction compressor	Compressor/sustainer for guitar
	Taurus	TUX MK2	Compressor/sustainer for guitar
	T-Rex Engineering	Tonebug sustainer	Compressor/sustainer for guitar
	Way Huge Electroniucs	Saffron squeeze	Compressor/sustainer for guitar
Rack	Alesis	3632	Compressor/limiter/expander/gate – 2 channels
	Aphex	CX-500	Compressor/noise gate – 1 channel
	ART	PRO VLA II	Optical tube compressor* – 2 channels
	Avalon	AD 2044	Optical compressor* – 2 channels
	Behringer	MDX2200	Compressor/limiter/gate – 2 channels
	Chandler Limited	RS124	Tube compressor – 1 channel
	Chandler Limited	Germanium compressor	Compressor – 1 channel
	DBX	160VU	Compressor – 1 channel
	DBX	266XL	Compressor/limiter/gate – 2 channels
	DBX	166XS	Compressor/limiter/gate – 2 channels
	Drawner	1968 MKII	Tube compressor – 2 channels

	Manufacturer	Model	Description
	Manley	Variable MU Mastering	Compressor/limiter – Tubes – 2 channels
	Fairchild	670	Compressor/limiter – Tubes – 2 channels
	NEVE	33609J/D	Compressor/limiter – 2 channels – Vintage
	Rupert Neve Designs	Portico 5043	Compressor/limiter – 2 channels
	Samson	S-Com plus	Compressor/limiter/expander/gate/ de-esser – 2 channels
	SSL	G Comp	Stereo compressor – API 500 Rackmount
	Tornade music systems	E-Series stereo bus compressor	Compressor – 2 channels – VCA**
	Universal Audio	LA2A	Optical compressor – 1 channel
	Urei JBL	1176LN	Compressor – 1 channel
	Vertigo Sound	VSC-2	Compressor – 2 channels – 4 VCA**
	Vintech Audio	609CA	Compressor/limiter – 2 channels
	Warm Audio	WA-2A	Tube compressor – 1 channel
	Weiss Engineering	Gambit DS1	Dynamic compressor/limiter – Wideband, multiband, and parallel – AES/EBU only
Software	Audiocation	Compressor AC-1	Freeware – Compressor Windows: VST
	Eventide	Omnipressor	Compressor/limiter/expander Windows: VST; RTAS; AAX OSX: AU; VST; RTAS; AAX
	IK Multimedia	Classic T-RackS Compressor	Compressor Windows: VST; RTAS; AAX OSX: AU; VST; RTAS; AAX
	MeldaProduction Audio Technologies	MCompressor	Freeware – Compressor/limiter Windows: VST; VST3; AAX OSX: AU; VST; VST3; AAX
	PSP Audioware	MasterComp	Compressor Windows: VST; RTAS; AAX OSX: AU; VST; RTAS; AAX
	Softube	Valley People Dyna-mite	Compressor/limiter/expander/de-esser/gate Windows: VST; VST3; RTAS; AAX OSX: AU; VST; VST3; RTAS; AAX
	URS	1975 Classic	Compressor/limiter Windows: VST; RTAS; AAX OSX: AU; VST; RTAS

Waves	H-Comp	Compressor Windows: VST; VST3; AAX; RTAS OSX: AU; VST; VST3; AAX; RTAS
Waves	PuigChild 670	Compressor – Emulation of the Fairchild 660 and 670 Windows: VST; VST3; AAX; RTAS OSX: AU; VST; VST3; AAX; RTAS
Waves	API 2500	Compressor Windows: VST; VST3; AAX; RTAS OSX: AU; VST; VST3; AAX; RTAS
Waves	SSL G-Master Buss Compressor	Compressor – Emulation of the G-Comp by SSL Windows: VST; VST3; AAX; RTAS OSX: AU; VST; VST3; AAX; RTAS
Waves	VComp	Compressor/limiter/de-esser – Emulation of the 2254 by Neve Windows: VST; VST3; AAX; RTAS OSX: AU; VST; VST3; AAX; RTAS
Waves	CLA-3A	Compressor/limiter – Emulation of the LA-3A by Universal Audio Windows: VST; VST3; AAX; RTAS OSX: AU; VST; VST3; AAX; RTAS
Waves	C1 Compressor	Compressor/expander/gate/equalizer Windows: VST; VST3; AAX; RTAS OSX: AU; VST; VST3; AAX; RTAS
Waves	Renaissance Compressor	Compressor Windows: VST; VST3; AAX; RTAS OSX: AU; VST; VST3; AAX; RTAS

* Optical compressors have a light source whose intensity varies according to the amount of signal passing through it and a photoelectric cell that reacts to the luminosity of the light source. These types of compressor can be either highly colored or very neutral. They provide an "organic" dimension to the sound.

**The VCA (voltage-controlled amplifier) is a transistor chip that monitors the level of the incoming signal to determine how much reduction needs to be applied. VCA compressors provide rapid and precise compression.

Table 7.1. *Examples of compression pedals, racks (studio) and plugins*

Studio compressors and pedals are based on several different technologies: optical compressors, field effect transistor (FET) compressors, tube compressors, voltage-controlled amplifier (VCA) compressors, operational transconductance amplifier compressors and many others. Each technology has its own pros and cons, and their performance in terms of sound rendering varies. To decide between them, you need to listen to how they sound within a given musical context. Most studios have multiple compressors to allow their sound engineers to add customized color or transparency to each mix.

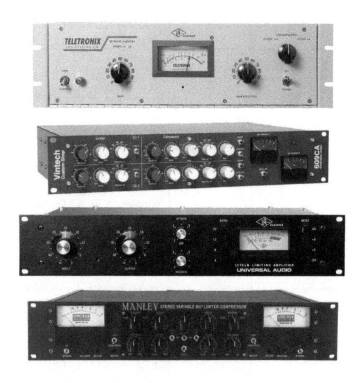

Figure 7.8. *Four well-known studio compressors (from top to bottom): LA-2A by Universal Audio (optical compressor), 609CA by Vintech and 1176 by Urei/JBL (FET compressors) and a Manley tube compressor*

There are a staggering number of software compressors – hundreds of them. They offer a broad spectrum of possibilities, as well as rich and very sophisticated settings. Many software compressors are clones of studio compressors.

7.1.4. Multiband compressors

This type of compressor separates the input signal into multiple bands (three or more) using an array of filters. Each band is then processed with a different compressor, before recombining the processed signals into a single output.

Dynamic Effects 167

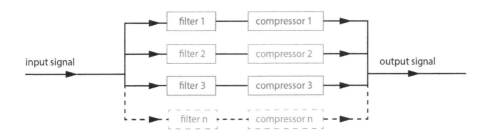

Figure 7.9. *Diagram of the principle of a multiband compressor. For a color version of this figure, see www.iste.co.uk/reveillac/soundeffects.zip*

These compressors can correct frequency imbalance in the original audio signal.

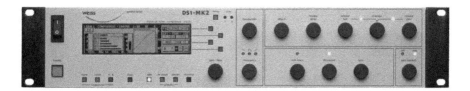

Figure 7.10. *The high-end digital dynamic studio rack compressor/limiter "Gambit DS1-MK2" by Weiis. This compressor is wideband, multiband and parallel (see section 7.1.6)*

Increasing the number of bands makes the compressor more complex. These types of processor are extremely delicate to operate, and dividing the signal into frequency bands is especially tricky.

Figure 7.11. *The multiband software compressor (3) Drawner 1973 by Softube*

Table 7.2 lists a few examples of multiband software compressors. There are many of them, so this list is far from exhaustive.

Publisher	Name	Remarks
Audio Damage	Rough Rider	Freeware – Multiband compressor Windows: VST OSX: AU; VST
Blue Cat Audio	MB-5 Dynamix	Multiband compressor Windows: VST; RTAS; AAX OSX: AU; VST; RTAS; AAX
Izotope	Ozone 7 – Vintage compressor	Multiband compressor Windows: RTAS; VST2; VST3; AAX OSX: RTAS; AU; VST2; VST3; AAX
Avid	Pro Tools	Multiband compressor Windows: AAX OSX: AAX
Nomad Factory	Essential Multiband Compressor	Multi-band compressor Windows: RTAS; VST OSX: RTAS; AU; VST
Nomad Factory	AMT Multi Max	Multiband compressor Windows: RTAS; VST OSX: RTAS; AU; VST
Softube	Drawner 1973 Multiband	Multiband compressor – Emulation of the Drawner 1973 compressor Windows: VST; VST3; AAX OSX: AU; VST; VST3; AAX
FabFilter	Pro-MB	Multiband compressor Windows: VST; VST3; AAX; RTAS OSX: AU; VST; VST3; AAX; RTAS
Waves	C6 Multiband Compressor	Multiband compressor Windows: VST; VST3; AAX; RTAS OSX: AU; VST; VST3; AAX; RTAS
Cakewalk	Sonitus	Multiband compressor Compatible with Sonar 4 Producer Edition and Sonar Studio Edition

Table 7.2. *Examples of multiband software compressors*

7.1.5. *Guidelines for configuring a compressor*

Properly configuring a compressor is extremely difficult. It is possibly one of the most complex tasks that you will encounter in mixing, and requires an enormous amount of subtlety and finesse. There is no magic recipe. It is a bit like cooking – even if you have the right ingredients, instructions and temperatures, you still need skill and experience to manage the vast number of unpredictable, incalculable or unknowable factors.

In the following, a sketch of the outlines of a simple method for studio applications is provided. As you gain experience, you can make your own refinements.

The objective is to stabilize the position of the sound signal relative to the position of the observer. We need to eliminate any sensation of back-and-forth motion:

1) Set the compression (threshold) to 0 dB;

2) Set the compression ratio to its maximum value;

3) Set the attack to its minimum value;

4) Set the release to its minimum value;

5) Begin playing the track;

6) Lower the threshold until the compressor is not constantly active;

7) Lower the compression ratio until the hypercompression caused by setting everything to max has been eliminated. At this point, you are looking to make the sound rendering more organic;

8) Increase the attack and listen to the effect on the transients; the perceived thickness of the sound should change;

9) Increase the release until any pumping is eliminated, letting the signal breathe. Look for the slowest possible setting that creates a groove and that you are happy with;

10) Adjust the knee until you find the right color and punch;

11) These settings should make the gain indicator dance around.

Remember, each parameter influences the others. You need to be tweaking and adjusting everything continuously, while listening carefully.

If you close your eyes, it should feel like the sound source has a stable position in space, with a characteristic groove and a natural quality.

If you lost gain during the processing, you can compensate by adjusting the increase/decrease setting. Some compressors have an automatic make-up gain function.

It is difficult to describe the sensations created by a compressor in writing. It truly is something that only your ears can fully appreciate. Do not hesitate to keep repeating your tests as many times as you need with your headset or via monitoring.

7.1.6. Parallel compression

Parallel compression is also known as "invisible compression" or "New York compression". It is a technique that makes sounds feel bigger without changing their dynamics.

It is often used for drums, but also works with other instruments, such as bass guitar or brass ensembles. It is not recommended for vocals.

The idea is to duplicate the track (for drums, first create a group that combines all percussions onto a single track). One of the new tracks is a copy of the original signal (optionally very slightly compressed), preserving the dynamics of the sound. The other track is compressed to obtain the sound matter.

Both tracks are then mixed into a single, definitive output signal.

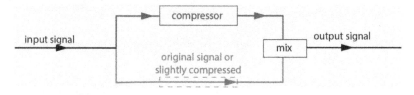

Figure 7.12. *Principle of parallel compression. For a color version of this figure, see www.iste.co.uk/reveillac/soundeffects.zip*

Be careful. It is very easy to end up with a sound that is confused and ugly. The key is to find the perfect balance, which in some cases can be extremely difficult.

One simple approach is to configure the compressor as follows: minimum threshold, maximum compression ratio, very quick attack, very short release and very hard knee.

Then play with the mix of original and compressed signals until you find a good compromise.

7.1.7. Serial compression

Just like electrical circuits, compression can be arranged in series as well as in parallel. With serial compression, the signal passes through two connected compressors consecutively. The goal is to obtain a powerful sound with lots of matter. These two objectives are contradictory with a single compressor.

The solution is to place multiple compressors in a certain predefined order. First, we process the sound matter and consistency, and then we give it power and muscle separately.

Each compressor has its own compression ratio. The overall compression ratio is obtained by multiplying the two individual ratios together, unlike the attack, release and knee, which are added together. This means that the sound rendering can become unnatural very quickly.

Serial compression is complex to implement. Much like parallel compression, it is very sensitive to its configuration. The right balance between the two compressors must be found in order to avoid an awful noisy mess.

7.1.8. Compression with a sidechain

We introduced the concept of sidechain in section 7.1.2.4. We now have a chance to go into a little more detail.

One of the possible applications of this technique is to prevent the instruments in a sound mix from masking each other. Instead of applying standard compression to the signal level, the compression is controlled by an external signal. For example, to clarify the sound rendering of a bass guitar, we can use the bass drum as an external source. Another example is to use the snare drum as an external source for vocals.

Sidechains are also extremely useful for synth pads. Due to their synthetic nature, synth pads often lack in dynamics. By using a strong, rhythmic component like the kick drum as a sidechain, we can give the synth pads more energy by varying their level with the rhythm.

This technique is called an *"external sidechain"*, as opposed to an *"internal sidechain"*, which we discussed previously.

We can also use the original signal itself as a trigger by compressing it and applying a high-pass or low-pass filter, allowing the compressor to target specific regions of the frequency spectrum. For drums, the lower end of the spectrum is often filtered to prevent the signal of the bass drum from being excessively compressed, which can drag the rest of the drums down with it, e.g. snare drum, tom-toms, etc. Doing this provides the sound rendering of the drums more clarity and transparency while still allowing compression to be achieved.

7.1.9. *Some basic compression settings*

This section gives guidelines for creating some of the various sounds that can be achieved with compressors, either for studio mixing or live sound reinforcement. However, there are no universal rules or methods. The following sections guide us in the right direction for good results.

7.1.9.1. *Improving sound cohesion*

Sometimes, a collection of multiple audio tracks requires extra coherence. Today, mixes can have many wildly different components, distributed over multiple distinct tracks, but which need to be linked to each other. Examples of such components might include recorded instruments, samples, electronic sounds, virtual instruments and so on. We can use a compressor to forge connections between components, either in combination with other methods or separately.

You will need the following settings:

– compressor with a well-defined color rather than a very transparent compressor. Vintage compressors or emulators should therefore be preferred;

– long attack time;

– suitable release time. If your compressor has an automatic mode, this is a good opportunity to use it;

– very soft knee;

– low compression ratio, between 1.1:1 and 2.0:1;

– threshold that reduces the signal by –0.5 to –1.5 dB.

7.1.9.2. *Increasing sound thickness*

Often, your sound recording or sequence will seem very or excessively clear, pure or discrete. You can reinforce it to occupy more space without losing its natural feel by increasing its thickness or consistency.

We need the following settings:

– long attack time for percussion instruments or any other rhythmic instruments, and short attack time for all other components;

– release time appropriate for the rhythm of the music. Automatic mode can be a good compromise, but is not always suitable; you will need to try it and see;

– knee as soft as possible;

– filtered bass frequencies to soften the compression;

– compression ratio as finely tuned as possible to retain the natural feel of the recording;

– threshold that grips (captures) each peak.

7.1.9.3. *Smoothening the sound*

Sometimes, you will want to create a very soft, fluid sound, with long attacks and little aggressiveness, without compromising on presence, naturalness or energy. This is often the case for the background of a mix, which supports the main sound sequence, or might for example be desirable for vocals in a slow part of a song.

You will need the following settings:

– very short attack time;

– rapid release;

– knee either soft or hard on a case-by-case basis. You will have to listen carefully to decide;

– high compression ratio, between 5:1 and 10:1;

– threshold that grips each peak;

– make-up gain to restore the signal to satisfactory levels. If your compressor has an automatic mode, you should use it, but it will not always be enough.

7.1.9.4. *Creating a powerful sound*

This technique places a sound at the front of the overall mix. It will seem more aggressive and present, closer to the audience and more detached from the other sounds. It will also seem less flattened and masked by other tracks that might cover parts of its frequency spectrum.

You will need the following settings:

– attack time long enough to preserve the natural timbre and feel of the signal;

– release time long enough to preserve the envelope without affecting the next group of transient frequencies;

– hard knee to ensure that the sound remains aggressive;

– high compression ratio to make the sound feel percussive without completely flattening it;

– threshold chosen to catch the most important peaks in the signal, namely those that carry and make up the dynamics;

– optionally, a make-up gain if you judge that the losses created by these settings need to be compensated.

7.1.9.5. Examples of standard settings

Table 7.3 lists a few standard settings that can be applied to individual tracks, either for recording or mixing.

Instrument	Compression ratio	Threshold	Attack	Release	Knee	Notes
Powerful drums	20:1 to max:1	5–10 dB	0.1–2 ms	40–100 ms	Hard	Usually applied to the bass drum
Drums	5:1	2–5 dB	1–10 ms	100–400 ms	Soft	
Bass guitar	4:1 to 10:1	3–6 dB	2–25 ms	250–350 ms	Hard	Longer attack for slap bass (harmonics)
Brass	4:1	5–8 dB for peaks	Auto	Auto		
Electric guitar	4:1 to 6:1	Depending on style of music, from 5–10 dB	5–30 ms	200–300 ms	Relatively hard, depending on style of music	Auto mode often works for attack and release
Acoustic guitar	3:1 to 4:1	3–5 dB	20–50 ms	200–300 ms	Soft	Try to preserve the natural sound of the instrument

Synthesizer	No compression, synth sound is usually already compressed enough					
Wind instruments	3:1	Relatively high, depending on instrument	5–10 ms	50–200 ms	Soft	Adjust depending on whether instrument plays solo
Studio vocals	2:1 to 6:1	5–12 dB for peaks	10–50 ms	100–400 ms	Soft	For rock music, reduce attack and sustain
Live vocals	2:1 to 4:1	5–10 dB for peaks	10–30 ms	100–300 ms	Soft	Auto release can be good for live sets
Solo voice	1.5:1 to 2:1	Max	10 ms	200 ms	Soft	Try to preserve the natural feel and grain of the voice

Table 7.3. *Examples of standard compression settings for specific instruments*

7.1.10. Synchronizing the compressor

If your compressor has a sidechain function with an external key input, there is an intriguing opportunity that you might like to explore.

You can plug a metronome or a track with the musical rhythm, like the hi-hat from the drums, into the key input.

Your compressor will synchronize with the external signal. This will enhance any instruments with weaker dynamics, like strings or synth pads.

In this way, you can create very special and original effects.

There is no hard-and-fast rule. You just need to play around, listen and then accept or reject the results depending on what you are looking for.

7.1.11. Using a compressor as a limiter

Theoretically, any compressor can act as a limiter if it allows the compression ratio to be set higher than 20:1, up to ∞:1 (which corresponds to a perfect limiter).

"Brickwall" limiters allow us to define a level or volume that will never be exceeded no matter what. This prevents clipping during mastering or on tracks with large differences in sound level.

Limiters tend not to have many configurable parameters. One parameter is the threshold, which defines the height of the limit, and another is the output level (or *ceiling*).

Some limiters also have a configurable release time, which defines the time taken by the limiter to stop acting on the signal, and thus the speed at which it does so.

Figure 7.13. *The software limiter L1 by Waves with its straightforward features (threshold, ceiling, release)*

Some software limiters have a few additional parameters.

Figure 7.14. *The software limiter LM-662 by Nomad Factory, a tool with more sophisticated parameters*

Limiters are often placed at the end of the processing chain in the mix (*main bus*), which is why software versions often include features for configuring the sampling depth (*dither* or *dithering*), which ranges from 8 to 24 bits.

Personally, I do not use limiters during mixing, because I find them too restrictive. But that is just me – some people do like to use them.

To use a limiter for mastering, you need to set the output level to around –0.5 to –0.2 dB. The threshold should only affect peaks, and its reduction level should never reach 0.

Table 7.4 lists a few examples of software limiters.

Publisher	Name or model	Remarks
Abbey Road	TG 12413 Limiter	Emulation of the EMI Abbey Road 1960 limiter – Limiter/compressor Windows: VST; RTAS OSX: AU; VST; RTAS
Blue Cat Audio	Blue Cat's Protector	"Brickwall" limiter Windows: VST; VST3; RTAS; AAX; DX OSX: AU; VST; VST3; RTAS; AAX
dB Audioware	dB-L Mastering Limiter 1.00	Freeware – Limiter Windows: DX
FabFilter	Pro-L	Multi-band compressor Windows: VST; VST3; RTAS; AAX OSX: AU; VST; VST3; RTAS; AAX
Kjaerhus Audio	Classic Master Limiter	Freeware – "Brickwall" limiter Windows: VST
MeldaProduction	MDynamics Limiter	Limiter Windows: VST; VST3; AAX OSX: AU; VST; VST3; AAX
Nomad Factory	BT Brickwall BW-2S	"Brickwall" limiter/compressor Windows: VST; RTAS OSX: VST; RTAS; AU
Nomad Factory	LM-662 Limiter	Emulation of the Fairchild 670 limiter Windows: VST; RTAS OSX: VST; RTAS; AU

| Sonnox | Oxford Limiter V2 | Limiter with variable knee
Windows: VST2; AAX; RTAS
OSX: VST2; RTAS; AU; AAX |
| Waves | L1 Limiter | Limiter
Windows: VST; VST3; RTAS; AAX
OSX: AU; VST; VST3; RTAS; AAX |

Table 7.4. *Examples of software limiters*

7.2. Expanders

This section briefly presents dynamic range expanders, which are included with most compressors and DAWs.

The purpose of an expander is to increase the dynamic range of an audio signal. Therefore, it operates like a reverse audio compressor.

7.2.1. Parameters

Above a certain threshold, the expander does nothing. Below this threshold, it applies a predefined expansion ratio.

This ratio takes values ranging from 1:2 and 1:4 up to 1:10, at which point the expander essentially behaves like a gate.

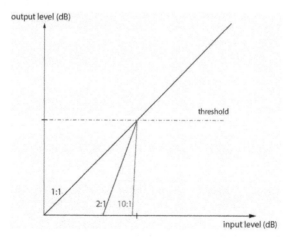

Figure 7.15. *Principle of an expander. For a color version of this figure, see www.iste.co.uk/reveillac/soundeffects.zip*

Just like a compressor, expanders have time parameters, including attack, release, knee and a *hold* function that defines the behavior of the expander after the signal exceeds the threshold, but before the release time begins.

Some expanders also have a sidechain mode, like most dynamics processing tools. Many compression plugins and studio compressors include an expander function.

7.2.2. *Examples of software expanders*

Table 7.5 lists a few expanders that are found in combination with noise gates (see below, section 7.3), compressors (see Table 7.1) and limiters (see section 7.1.11).

Type	Manufacturer or publisher	Model or name	Remarks
Rack	Behringer	Multigate Pro XR4400	Expander/gate – 4 channels
	Behringer	Multicom Pro-XL MDX4600	Expander/gate/compressor/limiter – 4 channels
	DBX	1074 QuadGate	Expander/gate – 4 channels
	Drawmer	DS101	Expander/gate – Rack unit, model 500
	Samson Technologies	S-Com plus	Compressor/gate/expander/de-esser/limiter
Software	Audio Damage	AD022 BigSeq-2	Noise gate/expander Windows: VST OSX: VST; AU
	FabFilter	Pro-G	Noise gate/expander Windows: VST; VST3; AAX; RTAS OSX: VST; VST3; AU; AAX; RTAS
	Kjaerhus Audio	GAG-1	Noise gate/expander Windows: VST OSX: VST; AU
	MeldaProduction	MStereoSpread	Noise gate/expander Windows: VST; VST3; AAX OSX: VST; VST3; AU; AAX

Nomad Factory	AS Expander	Gate	Noise gate/expander Windows: VST; AAX OSX: VST; AAX
Waves	Primary Source Expander		Noise gate/expander Windows: VST; VST3; AAX; RTAS OSX: VST; VST3; AU; AAX; RTAS

Table 7.5. *Examples of expanders*

7.3. Noise gates

As you may have guessed from their name, noise gates either block a sound signal or allow it to pass. They are used to eliminate any hiss or background noise on a track in a mix or from instruments, usually analog ones. An ideal noise gate does not modify the rendering or the color of the sound in any way.

Figure 7.16. *The DBX 1074, a studio noise gate*

This technique appeared in the late 1960s and was later adopted by high-end recording equipment in the 1980s for real-time applications.

7.3.1. *Parameters*

The parameters of noise gates are largely equivalent to those of compressors. Noise gates continuously analyze the signal, comparing it to a user-defined threshold. Whenever the overall level passes below this threshold, the noise gate closes, blocking the signal. The gate opens again once the level goes back over the threshold.

The attack and release parameters define the time taken by the gate to open and close whenever the signal crosses the threshold. The gate opens gradually during the attack to allow the audio signal to pass, and closes during the release until the signal is fully blocked.

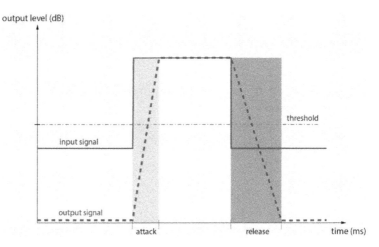

Figure 7.17. *Principle of the noise gate. For a color version of this figure, see www.iste.co.uk/reveillac/soundeffects.zip*

There can also be a *hold* parameter, which configures how long the gate should remain open after the signal falls below the threshold. The hold time elapses before the release begins (Figure 7.16).

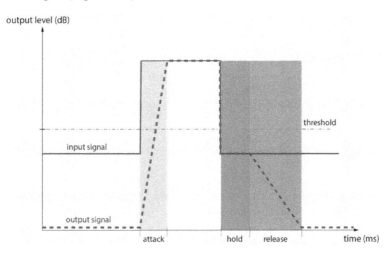

Figure 7.18. *Principle of a noise gate with a hold parameter. For a color version of this figure, see www.iste.co.uk/reveillac/soundeffects.zip*

The concept of hysteresis is often mentioned in connection with noise gates. The idea is to define two thresholds, the *open threshold* and the *close threshold*. The hysteresis is the difference between these values, which are measured in dB (Figure 7.17).

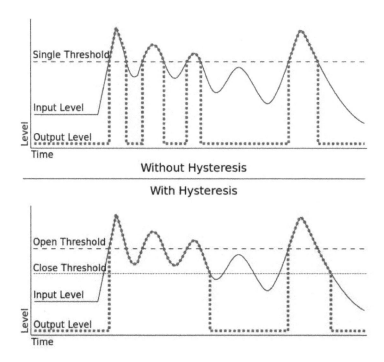

Figure 7.19. *Hysteresis in the context of an audio signal (input level) using a noise gate (top, without hysteresis; bottom, with hysteresis) (source: Wikipedia). For a color version of this figure, see www.iste.co.uk/reveillac/soundeffects.zip*

Some noise gates include an audio filtering system that only activates the gate on a certain frequency band.

Sidechains can also be used to control when the gate opens using an external signal. This can significantly improve the accuracy of the noise gate.

7.3.2. *Examples of noise gates*

Some examples of available software and hardware noise gates can be found in Table 7.5 and sections 7.1.3 and 7.2.2. Noise gates are often included with expanders or other dynamics processing equipment.

Dynamic Effects 183

Figure 7.20. *Four noise gate pedals (from left to right): "Noise Terminator" by Carl Martin; "The Silencer" by Electro Harmonix; "Sentry" by TC Electronic; "M195" by MXR*

Table 7.6 gives a few examples of noise gates pedals. These noise gates tend to be designed for guitar.

Brand	Name or model	Remarks
Boss	NS-2 Noise Suppressor	Available since 1987
Carl Martin	Noise Terminator	Two settings for two types of sound, hard and soft
Electro Harmonix	The Silencer	Three settings and an effects loop
Isp Technologies	Decimator II G-String	Noise reduction > 60 dB
MXR	M135	Very simple
MXR	M195	Very simple, with an effects loop
Rocktron	Hush The Pedal	Very simple, only one setting
Rocktron	Hush Super C	Rack
TC Electronic	Sentry	Multiband

Table 7.6. *Examples of noise gates for guitar*

7.3.3. Configuring noise gates

Noise gates are extremely simple to use. Simply set the threshold to the minimum value, then increase it until any undesirable signals have been eliminated.

You can then adjust the attack and release times to fine-tune the results.

If your noise gate has a filtering system, you will also need to configure the frequency band on which the noise gate will act. You can optionally configure an external trigger using the sidechain function.

7.4. De-essers

De-essers are a tool for reducing sibilance, hissing or squeaking sounds (e.g. soft "ch" and "s" sounds). These sounds are a natural part of vocals, but can cause problems in sound processing sequences that use compression, delay or reverb.

Figure 7.21. *The famous studio de-esser by SPL*

7.4.1. Principle of a de-esser

Before de-essers were available, compressors with an external input (key) connected to the sidechain function were used to solve this problem. This solution is still viable in some situations today.

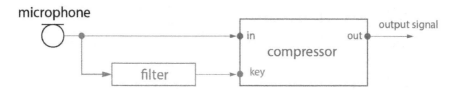

Figure 7.22. *Filter and compressor setup equivalent to a de-esser. The filter is often connected to an amp that increases sibilance to improve the compression. For a color version of this figure, see www.iste.co.uk/reveillac/soundeffects.zip*

However, de-essers work according to a slightly different principle. They only compress sibilance that was selected by their filters. The rest of the signal is left intact.

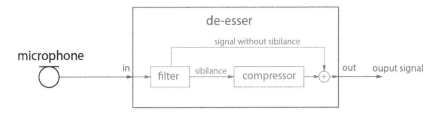

Figure 7.23. *Diagram showing the principle of a de-esser. For a color version of this figure, see www.iste.co.uk/reveillac/soundeffects.zip*

By default, de-essers are almost exclusively used as inserts on a track. They are straightforward to configure and often only have a (potentially automatic) threshold, frequency range or male/female preset and an attenuation range.

Some de-essers also have a frequency detector, attack and release settings, and various filters.

7.4.2. *Examples of de-essers*

Table 7.7 lists a few de-essers that you can find as studio racks or software plugins.

Type	Brand or publisher	Name or model	Remarks
Rack	Behringer	Pro-XL MDX2600	Expander/gate/compressor/de-esser/peak limiter/enhancer – 2 channels
	BSS Audio	DPR 422 Opal	Compressor/de-esser – 2 channels
	Dane	#31 Optical De-esser	1 channel
	DBX	520 De-esser	De-esser – Rack unit, model 500 – 1 channel
	DBX	263X	One channel
	Drawmer	MX50 – Dual de-esser	Two-channel de-esser
	Empirical Labs Deresser	EL-DS	One channel
	SPL	De-esser 9629	One of the most common studio de-essers – 2 channels
	Valley people	Micro FX	Expander/gate – 4 channels
Software	Digitalfishphones	Spitfish	Freeware Windows: VST

Eiosis	E²deesser	Windows: VST2; VST3; AAX
		OSX: VST2; VST3; AU; AAX
FabFilter	Pro-DS	Windows: VST; VST3; AAX; RTAS
		OSX: VST; VST3; AU; AAX; RTAS
Nomad Factory	BT Deesser DS-2S	Windows: VST; RTAS
		OSX: VST; AU; RTAS
Sonnox	Oxford SuprEsser	Windows: VST; RTAS
		OSX: VST; AU; RTAS
Waves	DeEsser	Windows: VST; VST3; AAX; RTAS
		OSX: VST; VST3; AU; AAX; RTAS
Waves	Renaissance DeEsser	Windows: VST; VST3; AAX; RTAS
		OSX: VST; VST3; AU; AAX; RTAS

Table 7.7. *Examples of de-esser racks and plugins*

7.4.3. Replacing a de-esser with an equalizer and a compressor

In the following, we explain a simple method for de-essing a vocals track with a studio mixing desk:

– send the vocals track that you want to process onto two channels of the desk;

– on the first channel, use the equalizer on the mixing desk to isolate the vocals as much as possible and amplify any sibilance;

– on the second channel, send the signal to a compressor;

– send the equalized signal to the key of the compressor's sidechain to eliminate the sibilance.

You can use the fader on the first channel to vary how much sibilance is eliminated, and the compressor on the second channel to control the de-essing rate.

7.4.4. Configuring a de-esser

Every de-esser is different, but you can adapt these basic settings to any model.

First, you need to determine the frequency at which your de-esser will be working. This is usually between 6 and 9 kHz (upper region of the spectrum). You need to narrow this down to reflect the vocal range of the singers: male, female, child, etc.

Next, configure the attenuation. This should usually be less than 8 dB to ensure that the sibilance is significantly reduced but that the dynamics of the signal are not flattened and that no gaps are created.

7.5. Saturation

Saturation in an analog audio circuit means that the strength of the processed signal cannot be increased no matter how much the input level is raised. Saturated signals exhibit *clipping*, a phenomenon that deforms the signal, also known as distortion.

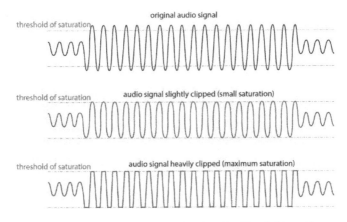

Figure 7.24. *Effect of saturation on an audio signal, with slight and heavy clipping. For a color version of this figure, see www.iste.co.uk/reveillac/soundeffects.zip*

In practice, clipping begins well below maximum amplification. Various types of intermediate effects can therefore be achieved, including *fuzz*, *overdrive* and *crunch*.

These effects strongly depend on the underlying technology: vacuum tubes, transistors, pre-amp equipment and any analog, digital or software systems.

This section presents some of the various types of saturation.

7.5.1. *Fuzz*

This effect creates a bold sound that is very rich in harmonics. It was one of the characteristic sounds of the 1960s and 1970s. This type of saturation has the most clipping, which tends to increase the sustain, i.e. the duration of each played note.

The effect depends strongly on the equipment (amps) and instruments. Many musicians have made it their own, and still use it to this day: Jimi Hendrix ("Purple Haze"), Carlos Santana, The Rolling Stones, Pink Floyd and many more. Legend has it that Johnny Burnette was one of the first to use it in 1957. Link Wray popularized the effect in 1958 with instrumental tracks like "Rumble" and, of course, "The Fuzz".

7.5.2. Overdrive

This effect imitates the sound of a tube amp pushed to maximum power, creating a very warm distortion that is highly popular with blues and rock guitarists (Chuck Berry, Ritchie Blackmore, Gary Moore, Steven Morse, Eddie Van Halen, etc.). Some organists have also used it, most notably with Hammond organs (Deep Purple, ELP, Pink Floyd, etc.).

Figure 7.25. *The model MA100H Marshall amp with overdrive settings (in the center of the control panel – magnified below)*

Some tube amps have an overdrive channel or setting that increases the signal gain of the instrument in the pre-amp phase, creating a saturated sound that is very rich in harmonics.

The overdrive effect is often also called "crunch" or "crunchy".

7.5.3. Distortion

Distortion is more aggressive than overdrive, offering a less natural and more clipped sound. Much like its close relatives, fuzz and overdrive, the sound of distortion strongly depends on the underlying equipment, and the various models that simulate it vary wildly from one another. Distortion can be aggressive, melodic, smooth, rich, soft, irregular, hard, creamy and much more.

REMARK.– Distortion, the effect used by musicians, is not quite the same as the more general concept of distortion, which is used in acoustics to describe audio signals. In the general sense, distortion refers to any alteration experienced by an audio signal as it travels through the processing and amplification chain including the final sound reproduction and playback system, e.g. speakers or headphones. In acoustics, distortion is characterized by several factors: the level, the *frequency response*, the *harmonic distortion*, the *cross-talk* and the *signal-to-noise ratio*. Appendix 1 explains this in more detail.

7.5.4. *Examples of equipment dedicated to creating saturation*

Pedals for generating fuzz, overdrive or distortion effects have proliferated since the 1970s. They simplify the process of configuring instruments and amplifiers, which is highly inconvenient on stage. The electronic components in the circuits used by saturation pedals have evolved over time, from germanium transistors with highly flawed stability and linearity to today's integrated circuits and software plugins.

Figure 7.26. *Three different types of pedal: "Big Muff" by Electro Harmonix (fuzz), "Tube Screamer" by Ibanez (overdrive) and "Distortion DS-1" by Boss (distortion)*

Table 7.8 lists a collection of pedals and other equipment (effects racks, software plugins) for creating and manipulating saturation effects. Most of these effects are designed for guitars, but they can also be used with other instruments.

A vast number of saturation pedals have been specifically designed for guitar – hundreds of them – and so this table only presents a very limited selection. The same is true for plugins.

Type	Brand or publisher	Model or name	Remarks
Pedals	Akai	Fuzz	Fuzz
	BBE	427	Distortion
	Boss	FZ-3	Fuzz
	Boss	ML-2	Distortion
	Boss	DS1-X Distortion	Distortion – Improved digital version of the DS-1
	Boss	DS-1 Distortion	Distortion
	Boss	BD-2	Overdrive
	Carl Martin	Classic DC Drive	Overdrive
	Carl Martin	Crush Zone Vintage	Distortion
	Digitech	Death Metal	Distortion
	Dunlop	JHF1	Fuzz
	Electro Harmonix	Big Muff	Fuzz
	Electro Harmonix	Micro metal Muff	Distortion
	Electro Harmonix	Bass Blogger	Distortion/overdrive/fuzz
	Fender	Blender Reissue	Fuzz
	Fender	Distortion	Distortion
	Fulltone	Octafuzz	Fuzz
	Ibanez	TS808 Tube Screamer	Overdrive
	Ibanez	DS7 Distortion	Distortion
	Line 6	Dr. Distorto	Distortion
	Maestro	MFZ-1	Fuzz
	Marshall	Shred Master	Distortion
	Mojo France	Bass Fuzz	Fuzz
	Mooer	Blues Grab	Overdrive
	Mooer	Fog	Distortion/overdrive/fuzz for bass guitar
	Morley	Distortion One	Distortion
	MXR	M236	Fuzz
	MXR	M75 Super Badass Distortion	Distortion
	MXR	M115 Distortion 3	Distortion
	MXR	M89	Distortion/overdrive/fuzz for bass guitar
	Rocktron	Rampage	Distortion
	Rocktron	Reaction Distortion 1	Distortion
	TC Electronic	MojoMojo	Overdrive
	TC Electronic	Vintage Distortion	Distortion
	TC Electronic	Dark Matter Distortion	Distortion

Software	T-Rex Engineering	Nitros	Distortion
	Audio Damage	FuzzPlus3	Freeware – Distortion/amplifier simulation Windows: VST2; VST3 OSX: AU; VST2; VST3
	AuraPlug	EddieVsHeaven	Freeware – Windows: VST OSX: AU; VST
	Blue Cat Audio	Destructor	Distortion/compression/amplifier simulation Windows: VST; VST3; AAX; RTAS OSX: VST; VST3; AAX; AU; RTAS
	BBE	Green Screamer	Distortion – Emulation of Ibanez Tube Screamer Windows: VST OSX: AU; VST
	FabFilter	Saturn	Distortion – Multiple styles – 150 presets Windows: VST; VST3; AAX; RTAS OSX: VST; VST3; AAX; AU; RTAS
	Izotope	Trash 2	Distortion – Multiband – Convolution and delay module – 300 presets Windows: VST; VST3; AAX; RTAS; DirectX OSX: VST; VST3; AAX; AU; RTAS
	MeldaProduction	MPolySaturator	Distortion – Choice of saturation spectrum Windows: VST; VST3; AAX OSX: VST; VST3; AAX; AU
	Softube	Acoustic feedback	Distortion/guitar feedback Windows: VST; VST3; AAX OSX: VST; VST3; AAX; AU
	SoundSpot	Halcyon	Distortion/overdrive Windows: VST; VST3; AAX OSX: VST; VST3; AAX; AU
	SoundToys	Decapitator	Distortion/overdrive – 5 presets

		Windows: VST; AAX
		OSX: VST; AAX; AU
SSL	X-Saturator	Distortion – Emulation of tube and transistor amps
		Windows: VST; VST3; RTAS; DirectX
		OSX: VST; VST3; AU; RTAS
Waves	OneKnob Driver	Overdrive
		Windows: VST; VST3; AAX; RTAS; DirectX
		OSX: VST; VST3; AAX; AU; RTAS
Wave Arts	Tube Saturator 2	Overdrive
		Windows: VST2; VST3; AAX
		OSX: AU; VST2; VST3; AAX

Table 7.8. *Examples of dedicated saturation effects*

7.6. Exciters and enhancers

Exciters, also called *aural exciters*, *harmonic exciters* and *enhancers* by some manufacturers, are devices that modify the audio signal by applying dynamic equalization, usually to the higher frequencies of the sound spectrum. Exciters can also be used to increase the harmonics of a sound with the goal of improving its contrast and therefore making it more intelligible.

Exciters can be difficult and sensitive to operate. They have the unfortunate tendency of caricaturizing sounds, especially vocals.

The original introduction of the famous "Aural Exciter" by Aphex in 1975 was met with fervent enthusiasm by many studios, who eagerly adopted it and immediately applied it to a vast number of their productions. Its popularity did not last. Exciters have a questionable effect on the quality, and sound technicians soon realized that it could easily be replaced with good equalization.

Figure 7.27. *The famous "Aural Exciter" by Aphex*

Since then, many other exciters have been created, some as racks, but most commonly as plugins. It is not easy to judge their appeal and benefits. Some users love them, others dislike them. You will need to try them out and listen to decide for yourself.

Figure 7.28. *Two modern rack enhancers: the "Sonic Maximizer 882i" by BBE and the "Exciter" model by Aphex*

You can also find exciter effects pedals, usually for guitar.

7.6.1. *Examples of exciters*

Table 7.9 gives a list of exciter pedals, studio racks and software plugins.

Type	Brand or publisher	Model or name	Remarks
Pedals	BBE	Sonic Stomp	-
	TC Electronic	Bodyrez	-
	Boss	EH-2	-
	Aphex	1401	Transistor pre-amp
	Aphex	Xciter	-
	Behringer	SE200 Spectrum Enhancer	-
Rack	Aphex	The Exciter	Two channels – XLR and jack inputs/outputs
	Aphex	204 Aural Exciter	Two channels – XLR and jack inputs/outputs
	Aphex	103A Aural Exciter	Two channels – Jack inputs/outputs
	Aphex	204 Aural Exciter	Two channels – XLR and jack inputs/outputs
	BBE	Sonic Maximizer 882i	Two channels – XLR and jack inputs/outputs – Input VU-meter with

	Behringer	Ultrafex II EX3100	LEDs and peak display
	Behringer	Ultrafex II EX3100	Two channels – XLR and jack inputs/outputs – Surround
	Behringer	Ultrafex Pro EX3200	Two channels – XLR and jack inputs/outputs – Surround
	Rocktron	RX20	Designed for guitar
	Rocktron	RX1	Designed for guitar
	SPL	Tube Vitalizer	Two channels – XLR and jack inputs/outputs – High-end exciter
	SPL	Spectralizer	99 presets – AES/EBU – S/P-DIF – RS422 – MIDI in/thru
	SPL	Vitalizer SX2 pro	Two channels – XLR and jack inputs/outputs
	SPL	Vitalizer Mk2-T	Two channels – XLR and jack inputs/outputs
Software	BBE	Sonic Sweet	Suite of several plugins: D82, H82, and L82 Windows: VST; RTAS OSX: VST; AU; RTAS
Software	Crysonic	Spectralive NXT-3	Windows: VST OSX: VST; AU
Software	MeldaProduction	MStereoProcessor	Windows: VST2; VST3; AAX OSX: VST2; VST3; AU; AAX
Software	Nomad Factory	AS-Exciter	Windows: VST; AAX; RTAS OSX: VST; AAX; RTAS
Software	SPL	Vitalizer MK2-T	Windows: VST2; VST3; AAX OSX: VST2; VST3; AU; AAX
Software	Voxengo	Warmifier	Windows: VST; VST3 OSX: VST; VST3; AU
Software	Waves	Renaissance Bass	Windows: VST; VST3; AAX; RTAS OSX: VST; VST3; AU; AAX; RTAS
Software	Waves	Maserati B72	Windows: VST; VST3; AAX; RTAS OSX: VST; VST3; AU; AAX; RTAS
Software	Waves	Vitamin	Windows: VST; VST3; AAX; RTAS OSX: VST; VST3; AU; AAX; RTAS

Table 7.9. *Examples of exciters*

7.6.2. *Using a sound enhancer*

Exciters might seem like impressive tools that enhance the substance of a sound message by creating a richer, warmer signal with increased presence.

However, you need to be careful and apply them sparingly. They create harmonic distortion, which can quickly become uncomfortable for the audience.

Try to avoid processing the full bandwidth of the signal. Only process the bands of the spectrum that really need it, based on any issues that you noticed by listening.

You need to bear in mind that your ear will quickly get used to the extra sound information added by the enhancer. You need to be very careful and make adjustments in small increments.

Take frequent rests from your work, allowing a few hours for your hearing to settle so that your ear can recover its selectiveness and regain the ability to discriminate sufficiently well between signals.

7.7. Conclusion

Dynamic effects can really improve sounds, but they are extremely subtle. Experience is required to achieve a good rendering, which can all too easily spiral into an awful mess that is unpleasant to listen to.

We need to remain alert. Always listen, then listen again before doing something, and do not hesitate to revert any changes that do not work. Remember that it is often better to use as few effects as possible to preserve the substance of the original sound recording, assuming that the recording was properly executed in favorable conditions.

The same advice applies to sound reinforcement and live sets. Keep things simple and relaxed – just a little is often enough, and too much is too much.

8

Time Effects

The category of effects studied in this chapter revolves around time, which is one of the most important parameters for reverb and delay. Although the speed of sound (around 340 m/s[1]) might seem high at first glance, humans are actually capable of perceiving and interpreting tiny and subtle time variations in sound messages when they listen carefully.

Time-based effects are sometimes also called *spatialization* effects, since they perfectly characterize the sound space of a room or hall.

8.1. Reverb

Reverb is without a doubt one of the most commonly encountered effects in modern recordings. Since reverb arises naturally in some environments but can also be created artificially, it attracts a great deal of controversy. Purists claim that artificial reverb is quite simply wrong. Many others view reverb as a highly relevant and fully configurable creative tool that offers an infinite range of possibilities to musicians and sound engineers, allowing them to redefine and remodel the ambience of their sound palette.

8.1.1. *Theoretical principles*

As sound waves propagate through the air, they are reflected by any obstacles, walls, ground, ceiling or other objects that they encounter. A single wave can be reflected multiple times, causing it to follow longer and longer paths. As it does so, its energy gradually decreases, absorbed by the reflecting materials.

---

1 Rate of propagation of sound in air with normal temperature and pressure conditions.

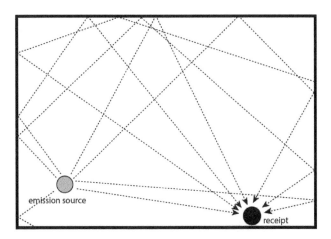

Figure 8.1. *Examples of possible trajectories (dotted lines) for the reflections of a sound wave in a room, creating reverberation*

The number of reflections is usually large, since the speed of sound (around 340 m/s) is typically large relative to the size of the location where the sound is propagating. In a room that measures 10 m by 10 m, we can expect around 30 reflections per second.

Observers in a reverberating environment will hear the same sound signal repeatedly at slightly different times. This stretches the sound, making it last for longer than the sound that was originally emitted. This phenomenon is known as reverberation (reverb), and the additional duration by which the sound signal is extended is called the *reverb time*.

Reverb often makes sounds seem more lingering (large spaces like a church, cathedral, etc.) and/or louder (small venues without much sound absorption) than sounds that are played outdoors.

In small rooms, the time differences are small enough that the sound signal does not repeat itself, and the observer perceives it as a single sound. If the dimensions of the space in which the sound is emitted are sufficiently large (like a valley) or are concave shaped (like a tunnel), the sound signal takes much long to return to the observer, who can then distinguish between each separate instance (repetitions). This is known as an *echo*, which is a special case of reverb.

In order to measure reverb, the field of acoustics defines the theoretical reverb time of a given location as the time taken for the sound energy to drop to one-millionth of its initial value, which corresponds to an attenuation of 60 dB.

Architects often analyze the reverb time when designing buildings. Sabine's equation[2] can be used to calculate the theoretical reverb time.

$$T = \frac{0.163 \times V}{A}$$

where:

- 0.163: constant in s/m;
- T: reverb time in s;
- V: volume of the room in m^3;
- A: absorption surface area in m^2 (sum of the product of the area of each surface S_n in m^2 and its Sabine absorption coefficient α_n).

$$A = (\alpha_1 \times S_1) + (\alpha_2 \times S_2) + \cdots$$

Table 8.1 gives a few examples of Sabine absorption coefficients as a function of the sound wave frequency.

Material	125 Hz	250 Hz	500 Hz	1 kHz	2 kHz	5 kHz
Wood	0.09	0.11	0.1	0.08	0.08	0.1
Cement coating	0.01	0.01	0.02	0.02	0.02	0.03
Wallpaper	0.01	0.02	0.04	0.1	0.2	0.3
Carpet	0.12	0.2	0.25	0.45	0.4	0.35
4-cm-thick rock wool	0.3	0.7	0.88	0.85	0.65	0.6
Exposed plaster	0.04	0.03	0.03	0.04	0.05	0.08
Painted plaster	0.01	0.01	0.02	0.03	0.04	0.05
Exposed concrete	0.01	0.01	0.01	0.02	0.05	0.07
5 mm plywood + 5 cm air	0.47	0.34	0.3	0.11	0.08	0.08
Tiles	0.01	0.015	0.02	0.025	0.03	0.04
Ordinary glazing	0.3	0.22	0.17	0.14	0.1	0.02

Table 8.1. *Sabine absorption coefficients of various materials at different frequencies*

---

2 Wallace Clement Sabine, born in Colombus, OH, USA, 1868–1919. Acoustic engineer and Harvard graduate.

As a_n tends to 1, the wall absorbs an increasing proportion of the sound energy. If no energy is reflected, the material is said to be perfectly absorbent. As a_n tends to 0, the material becomes increasingly reverberant.

Sabine also showed (Sabine's laws):

– the curves describing the build-up and decay of a sound are essentially exponential, and are complementary, meaning that the increase in sound energy over a given period is equal to the decrease over the same period;

– the effect of an absorbent material does not depend on its position;

– the release period of a sound is approximately the same at every point in the sound space;

– the reverb is independent of the position of the sound source within the space.

However, these laws need to be nuanced in some cases. For example, reverb works slightly different in the presence of obstacles, people and convex or concave walls.

8.1.2. *History*

Reverberation was integrated into early recordings by physically performing the recording in a sound space with natural reverb.

In the early 1920s, the first record companies actively hunted for good locations and venues for recording reverb. Their sound engineers devised various ingenuous strategies for arranging their microphones to create specific effects.

The engineer Bill Fine was one of the first to use natural reverb in sound recordings. He simply used a single microphone to capture the ambient sound of the room. His recordings were produced and published by Mercury Records on a disc in their "Living Presence" series.

The music industry later began to design and build dedicated acoustic reverb chambers. These chambers were typically closed spaces whose walls had non-parallel surfaces and shapes, coated with shellac to make them highly reflective. These shapes, often called *diffusors*, were designed to break up the reflections and return them chaotically. A speaker and one or several microphones were arranged around the room to record the reflections of the original sound.

Figure 8.2. *One of the reverb chambers at Abbey Road Studios. You can see the cylindrical diffusors at the back of the room. The speaker that plays the sound can be seen in the foreground (source: CBS Interactive)*

Improvements in recording techniques, microphones and the invention of "high-fidelity[3]" after the war caused reverb to gain in popularity. Today, it has become one of the most classic effects.

Figure 8.3. *Acoustics Research Society Dresden, reverb chamber with multiple diffusors (source: wikiwand)*

---

3 Also known as "hi-fi". This term is used to describe audio equipment that is of higher quality than conventional equipment. The goal is to achieve sound reproduction that recreates the original sound as closely as possible for audiophiles.

One of the first creative applications of reverb was proposed by Bill Putnam[4], the founder of the first independent recording studio, Universal Recording, Chicago, USA, in 1947. He incorporated the natural reverb of his studio bathroom into the recording of the song "Peg O' My Heart[5]" by the Harmonicats. The result was hugely successful.

Figure 8.4. *"Peg O' My Heart" by the "Harmonicats"*

Over the next few years, like many other large studios, he would commission the construction of dedicated rooms of various sizes to create beautiful reverb effects. At the time, this was the only way of obtaining artificial reverb. It would continue to be widely used by recording studios during the 1960s.

REMARK.– Many reverb chambers still exist today, often for research purposes.

By changing how the recording microphone(s), the speaker and the diffusors are arranged, sound engineers can create very different reverb effects within the same chamber.

---

4 American sound engineer, author, publisher, producer, 1920–1989. He is viewed as the father of modern sound recording.

5 You can listen to this song on YouTube at the following address: https://youtu.be/9BIuX7IsdE8.

Although reverb chambers are highly effective, it was difficult to achieve long reverb times similar to those found in a cathedral, as this requires enormous spaces.

Because of technological progress, one of the first solutions to this problem was developed in 1957 by the German company EMT[6] (ElektroMessTecknik). They made a truly innovative contribution by developing and marketing a *plate reverberation unit*, the EMT 140, developed in collaboration with the Nuremberg Technical Broadcasting Institute and the Hamburg Institute of Radio Engineering, the workplace of Walter Kuhl, the engineer who invented it.

The EMT 140 is excellent at creating reverb, and is still used today in many studios. However, it has a major disadvantage: its bulky size of 2.30 × 1.25 × 0.31 m and its weight of more than 170 kg.

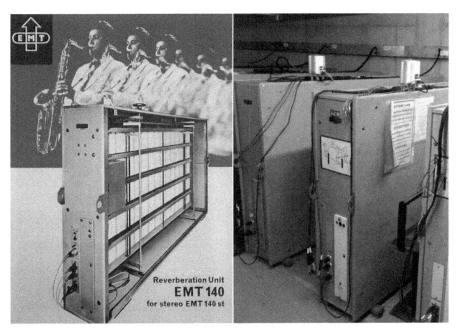

Figure 8.5. *The service manual of the EMT 140 (left) and several operational EMT 140 units (right) (Source: Basta Music)*

In 1959, another less cumbersome solution was invented by Laurens Hammond, the inventor of the famous organ bearing his name. Since 1935, Hammond organ

---

6 Today, EMT is considered the oldest and most popular manufacturer of reverb solutions. They have made constant improvements to their products over the years.

owners had been eager to recreate the sound of the pipe organs found in churches and theaters. However, this had proven impossible so far, since this kind of reverb requires very particular architecture.

Hammond, like any good inventor and businessman, was interested in meeting this demand for reverb equipment. After studying the work conducted by Bell Laboratories, he discovered that they had invented a spring-based electromechanical device that was capable of simulating delays, which they were currently using for long-distance telephone calls. By modifying this device, Hammond developed a satisfactory artificial reverb system. The first version of this system was around 1.20 m in height.

Hammond's system was then improved by the three engineers Alan Young, Herbert Canfield and Bert Meinema, who developed so-called "necklace" reverb. The name is inspired by the way that the springs in the device hang like a necklace (see Figure 8.6). It was much smaller than the original system: about 35 cm wide, 2.5 cm thick and 36 cm tall.

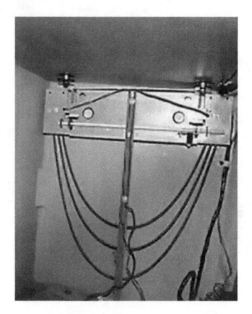

Figure 8.6. *A reverb "necklace" placed inside the amplification cabinet of a Hammond organ*

This reverb system was further improved and made smaller in the early 1960s (Figure 8.7).

Figure 8.7. *The famous Hammond spring reverb, Accutronics type 4 model. This system became an industry standard*

It is still widely manufactured and used today, especially inside various instrument amps by brands such as Fender, Marshall, Peavey, Ampeg and many others.

In 1976, at the AES convention in Zurich, Karl Otto Bäder, the technical director of EMT, revolutionized the reverb industry by unveiling the first ever digital reverb, the EMT 250 (Figure 8.8). This represented a huge step forwards, reinvigorating reverb with an impressive range of possibilities given the technology of the time. However, it was prohibitively expensive, costing almost $15,000. Even today, the EMT 250 arguably still fully holds its own against the latest generations of reverb hardware.

Figure 8.8. *The famous EMT 250 digital reverb*

Two years later, the American company Lexicon unveiled the model 224 for less than half the price. This model would become a staple of recording studios worldwide.

I have deliberately avoided talking about so-called acoustic tube or duct reverb and magnetic tape reverb. We will discuss these types of reverb later in section 8.1.3.5.

Figure 8.9. *The Lexicon 224 digital reverb*

Today, Lexicon is synonymous with reverb for sound professionals. The Lexicon brand has forged an outstanding presence in every domain of sound effects.

The model 224 was followed by the model 480L in 1986, which offered several innovative features, including an implementation of MIDI (Musical Instrument Digital Interface). The 480 range was extended by various new additions over time, from the 1.01 to the 4.01, before ultimately being superseded by the latest generation, the 960L.

Figure 8.10. *The 480L reverb unit by Lexicon*

In 1999, Sony introduced a real-time convolution reverb processor named the DRE S777. This was another massive technical advancement that heralded another round of improvements to reverb systems.

Figure 8.11. *The SONY DRE-S777 reverb effects processor*

Since the early 2000s, many distributors and manufacturers have marketed reverb equipment (or reverb processors) of various calibers and price brackets: Lexicon, TC Electronic, Alesis and many more.

In parallel, software reverb began to emerge in the form of plugins. One of the earliest and the best-known plugins is the famous Altiverb (*convolution reverb*) developed by the Dutch company Audio Ease. It would be followed by many others.

Figure 8.12. *The Altiverb reverb plugin by Audio Ease*

REMARK.– Software reverb has not rendered other types of reverb obsolete.

8.1.3. *Principles*

In this section, we will discuss the main types of reverb technology that we encountered in section 8.1.2. We will discover how each of them works.

We will study them in the following order:

– analog plate reverb;

– analog spring reverb;

– digital reverb based on algorithms;

– digital convolution reverb;

– other types of reverb.

8.1.3.1. *Analog plate reverberation*

This type of reverb is based on a metal plate, usually a steel sheet (the EMT 240 model uses gold leaf and was a lot smaller than its competitor, the EMT 140, but this meant that it was extremely fragile – see Figure 8.13).

This plate (1 × 2 m sheet 0.5 mm thick in the case of the EMT 140) is stretched taut over a metal frame mounted in a wooden box to protect it from the outside environment. After amplification, a *transducer* in the center of the plate transmits the sound by vibrating the sheet, transforming the original signal into a mechanical signal (wave) that travels through the plate and is partially reflected. A sensor (or two sensors for stereo) placed at one (or both) end(s) records the reverberated sound.

Figure 8.13. *The EMT 240 (produced in 1972), whose plate includes a 30 × 30 cm gold sheet that is 0.018 mm thick. Its outer dimensions are 64 × 30 × 62.5 cm, and it weighs 67 kg*

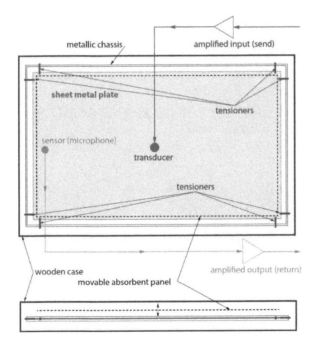

Figure 8.14. *Principle of plate reverberation. For a color version of this figure, see www.iste.co.uk/reveillac/soundeffects.zip*

The duration of the reverb (up to 5 s) is controlled by a parallel plate made up of absorbent material whose distance from the sheet can be adjusted. The closer the parallel plate is placed, the greater the damping, and the shorter the reverb time.

The system that moves the absorbent panel to modify the reverb duration can be motorized to allow it to be controlled remotely.

Figure 8.15. *Two mechanisms for controlling the reverb time: manual (left) and motorized (right)*

A more recent version of the EMT 140, known as the EMT 140 TS and released in 1971, supports stereophonic sound.

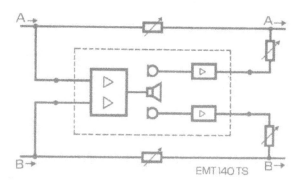

Figure 8.16. *Diagram of the amplification system of the EMT 140 TS (source: EMT)*

The EMT 140 TS boasts a range of improvements, including a double transducer and two sensors. Furthermore, its electronics (EMT 162 TS amplifier) were more sophisticated using a transistor-based system instead of electronic tubes.

8.1.3.2. Analog spring reverb

The principle of this reverb is equivalent to plate reverb. After being amplified, the signal is transmitted to a transducer that excites the end of one or several springs. At the other end, a sensor recovers the mechanical motion, reconverting it into an electronic signal, then sends it onwards.

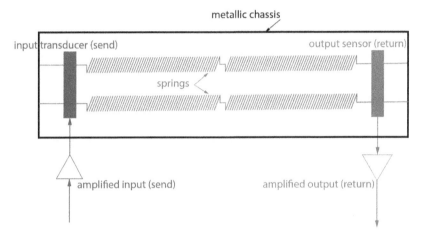

Figure 8.17. *Principle of spring reverberation. For a color version of this figure, see www.iste.co.uk/reveillac/soundeffects.zip*

When the sound wave reaches one end of the spring, part of its kinetic energy is reflected and redirected into the coils. These reflections create the reverberation.

Figure 8.18. *Close-up of the sensors and transducers. You can see the springs, the coil and its air gap, and the two magnets (right) in between. For a color version of this figure, see www.iste.co.uk/reveillac/soundeffects.zip*

The springs, the transducer and the sensor are usually mounted in an open metal case fixed to the amplifier or other system.

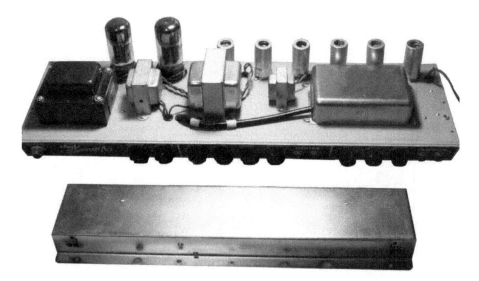

Figure 8.19. *The electronics of the "Vibraverb" amplifier by Fender. This was one of the first amps (1963) to incorporate spring reverb (bottom image)*

8.1.3.3. *Algorithm-based digital reverb*

This type of reverb is also described as synthetic reverb. It uses algorithms to model sounds by simulating a space with specific shape, size and other properties (damping, surface type, materials, etc.).

One helpful analogy to describe the way that algorithmic reverb generates effects is to think about how synthesizers create the musical sound of instruments (flute, oboe, organ, piano, violin, etc.).

Algorithmic reverb uses delay lines based on specialized circuits dedicated to *DSP* (digital signal processing). These circuits apply special algorithms such as circular buffering[7].

After being attenuated, the delayed signals are mixed with the original signal to achieve the desired effect. The number of delayed signals can be relatively high. The

---

[7] The circular buffering algorithm uses dedicated DSP registers called DAGs (Data Address Generators).

first signals represent the primary reflections of the sound, the next represent reflections of reflections and so on. This creates a *densification* phenomenon in the sound message, which becomes more diffuse. This diffusion represents and characterizes the intrinsic nature of the reverb effect.

This explanation is very simplistic. In reality, things are much more complex, since reverb behaves differently depending on the nature of the sound signal: pulses, continuous, sequential; and its frequency: treble, mid or bass.

Effects developers and manufacturers need to balance all of these factors in order to create high-quality reverb that is as faithful as possible, depending on what their users want.

To find out more about signal processing and the complexity of audio processing algorithms, you can visit the relevant entries in the bibliography and the web links at the end of this book.

Synthetic reverb can be found in the form of hardware, such as pedals or studio processors, sometimes combined with other effects, or in the form of software plugins. It would be very difficult to list all of them. Table 8.2 gives a non-exhaustive list[8] of synthetic reverb effects classified by category and manufacturer.

Type	Manufacturer or publisher	Name or model	Remarks
Pedal	Boss	RV-5	Six types of reverb – Stereo
		RV-6	Eight types of reverb – Stereo
	Electro Harmonix	Holy Grail	Three types of reverb
		Cathedral	Eight types of reverb – Stereo
	Eventide	TimeFactor	Multieffects – 10 types of effect (echo, reverb, delays, etc.) – Stereo
	MXR	M300	Six types of reverb
	TC Electronic	Alter Ego V2	Multieffects – 9 types of effect
		Arena reverb	Multieffects – 10 types of effect
Rack	Alesis	MicroVerb 3	Sixteen types of effect – Stereo effects (reverb, delay)
		MicroVerb 4	Multieffects (reverb, delay, chorus, etc.) – 100 presets – 100 user programs
		NanoVerb 2	Multieffects (reverb, delay, chorus, etc.) – 16 presets – Stereo
	Eventide	Reverb 2016	High-end reverb – 99 user presets – Stereo

---

8 For more information on the characteristics of each model of reverb, including sales price, manufacturing date, etc., and to find more models, you can visit the corresponding section of the web links at the end of the book.

	Lexicon	PCM 81	High-end multieffects – 300 presets – Stereo
		PCM 96	High-end reverb – 300 presets – Stereo
		MX 200	Reverb – 99 presets – 99 user programs – Stereo
		MX 500	Reverb – 99 presets – 99 user programs – Stereo
		MPX-1	Multieffects – 200 presets – 50 user programs – Stereo
		MPX 550	Multieffects – 240 presets – 64 user programs – Stereo
	TC Electronic	M3000	High-end multieffects – 250 simple presets + 50 combined presets – 250 user presets + 50 combined presets
		Mastering 6000	Very high end multichannel reverb and mastering processor
Software	Waves	H-Reverb	Windows: AAX; VST; VST3 / OSX: AAX; AU; VST; VST3
	UVI	Sparkverb	Windows: AAX; VST / OSX: AU; AAX; VST
	Lexicon	MPX/PCM/LXP	Windows: RTAS; VST / OSX: RTAS; AU; VST
	LiquidSonics	Seventh Heaven	Windows: VST2; VST3; AAX / OSX: VST2; VST3; AAX; AU
	PSP AudioWare	Nexcellence	Windows: RTAS; VST2; VST3; AAX / OSX: RTAS; AAX; AU; VST2; VST3
	Rob Papen	RP-Verb 2	Windows: VST; AAX / OSX: VST; AAX; AU
	Eventide	UltraReverb/TVerb	Windows: AAX; VST / OSX: AU; AAX; VST
	Universal Audio	EMT 140	Windows: RTAS; VST; AAX / OSX: RTAS; AAX; AU; VST / UAD-1 and UAD-2 cards
	Valhalla DSP	ValhallaRoom ValhallaPlate	Windows: RTAS; VST; AAX / OSX: RTAS; AAX; AU; VST
	Sinusweb	FreeverbToo	Freeware – Windows: VST; DX
	Kresearch	KR-Reverb FSQ	Freeware / Windows: VST; AAX / OSX: VST; AAX; AU
	Rhythm Lab	Mo'Verb	Freeware – Windows: VST
	TC Electronic	M30/ClassicVerb/DVR2 powercore	Windows: VST / OSX: VST; AU
	TC Electronic	VSS3	Windows: VST2.4; VST3; AAX / OSX: VST2.4; VST3; AAX; AU

Table 8.2. *Examples of algorithmic reverb effects by type*

8.1.3.4. *Digital convolution reverb*

This is the most modern type of reverb. Convolution reverb works by sampling the sound of a room. The system records the reflections, then analyzes them to build an accurate model of the room that includes a description of its shape and its properties.

As an analogy, you can think of a musical sampler, which records samples of the sound created by a musical instrument in order to be able to recreate that instrument.

In the following few sections, I will give a simplified presentation of the concept of convolution in the context of sound processing. For more details, and for discussions on this subject from a deeper mathematical perspective, you can visit the bibliography and web links at the end of the book.

Convolution is a mathematical operation on two functions. These functions are often denoted $x(t)$ and $h(t)$. The first defines the input signal, and the second defines the impulse response (IR).

What is the IR? It is an acoustic signature that characterizes a given location. The IR is determined by recording a sound message using one of two possible methods: the *free-field* method, also known as the "*transients*" method, and the "*sine sweep*" method.

The first method typically records a very short signal that contains every audible frequency with no echo, such as a balloon bursting, gunfire, etc.

The second method records a "sweeping" sound, which is simply a sine wave that ranges over every frequency within a relatively short period of time (a few seconds).

Once the recording is complete, *deconvolution*[9] is performed, which subtracts the original known signal to isolate the IR.

The function $y(t)$ is the convolution product, i.e. the result of the convolution $x(t) \times h(t)$. This is the output signal obtained by applying the IR to the sound message, which adds reverb to it.

In summary, we calculate the sound signature of a location, and then we process any sound sequence that we like in order to endow it with the characteristics of the reverb encoded in this signature.

---

9 Deconvolution is not required for the signals used by the free-field method, since these signals are a very close approximation of a Dirac pulse. A Dirac pulse is a signal with zero duration and infinite amplitude. This type of source signal is used in the mathematical theory of convolutions.

Convolution reverb effects are very faithful, convincing and high quality, but are limited by the rigidity of their implementation (based on a static recording). They provide very limited options for configuring the sound rendering. However, significant progress has been made more recently, and the latest generations allow attenuation, equalization and spatialization to be configured, as well as offering a "reverse" mode.

Table 8.3 lists a few examples of convolution reverb[10]. They are all software based (plugins). This effect no longer exists in the form of studio processors (racks).

Publisher	Name	Remarks
Acustica Audio	Nebula HS Reverb	– Windows: VST – (freeware)
Audio Ease	Altiverb 6 XL	– Windows: AAX; VST; RTAS – OSX: AAX; AU; MAS; VST; RTAS; TDM
Audio Ease	Altiverb 7	– Windows: AAX; VST – OSX: AAX; AU; MAS; VST; RTAS
Audio Ease	Altiverb 7 XL	– Windows: AAX; VST; RTAS – OSX: AAX; AU; MAS; VST; RTAS; TDM
Native instrument	Reflektor	– Windows – OSX
Thomas Resch	Tconvolution	– OSX (freeware): VST; AU; RTAS
VSL	Vienna MIR	– Windows: VST; AAX; RTAS – OSX: VST; AAX; RTAS; AU
VSL	Vienna MIR Pro	– Windows: VST; AAX; RTAS – OSX: VST; AAX; RTAS; AU
Waves	IR360	– Windows: VST; AAX; RTAS – OSX: VST; AAX; RTAS; AU

Table 8.3. *Examples of convolution reverb*

8.1.3.5. *Other types of reverb*

As promised earlier, we will now discuss the so-called acoustic tube or duct reverb, as well as magnetic tape and magnetic plate reverb.

The latter types of reverb are actually more similar to echo chambers or delay effects, although they can sometimes also create the illusion of reverberation.

---

10 For more information on the characteristics of each model of reverb, including sales price, manufacturing date, etc., and to find more models, you can visit the corresponding section of the web links at the end of this book.

8.1.3.5.1. Acoustic duct

Acoustic duct reverb is created using a long tube with a speaker at one end and a microphone at the other. The speaker plays the sound message, and the microphone records it.

Of course, to achieve a noticeable delay, the tube needs to be very long, because the speed of sound is relatively high (approximately 340 m/s).

For a 17-m tube:

$17/340 = 0.05$ s $= 50$ ms delay

For a 51-m tube:

$51/340 = 0.15$ s $= 150$ ms delay

The duct can be bent like an accordion to save some space, at the cost of some loss in sound quality, which is already not great with this technique. The bandwidth is limited and modified by the nature of the material from which the duct is made, as well as its width, shape and the properties of the transmitter (speaker) and receiver (microphone).

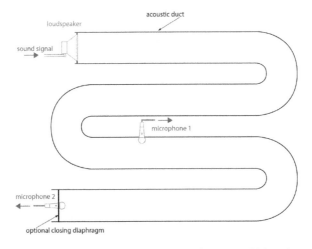

Figure 8.20. *Principle of an acoustic duct. By placing multiple microphones at different points along the duct, we obtain reverbs with different durations. For a color version of this figure, see www.iste.co.uk/reveillac/soundeffects.zip*

If the tube is very long, multiple microphones can be placed at different positions to obtain shorter and longer reverb times. The duct can either be open or closed.

8.1.3.5.2. Magnetic tapes or plates

This type of reverb is created using either a magnetic tape loop or a circular magnetic plate.

It works as follows:

– the audio signal input is recorded on the tape or plate by a write head placed at a certain position;

– a read head is placed after the write head in the direction of scrolling/rotation. This second head reads the signal, then simultaneously sends it to the output while also feeding it back to the write head;

– one or multiple other read heads can similarly be placed at various distances behind the first.

If the write head and the read head are positioned close together, or if the tape/plate is scrolling/rotating quickly, the repetitions will overlap, and will be indistinguishable to the naked ear.

Conversely, if the scrolling/rotation speed is slow and/or the heads are far apart, the reverb becomes an echo or a delay. We will revisit these concepts later in section 8.2.

Figure 8.21. *The famous Roland echo chamber "Space Echo RE-201" manufactured between 1974 and 1990. Its looping magnetic tape system is visible in the bottom left image, and its write head and four read heads can be seen in the bottom right image*

REMARK.– The part of the signal that is fed back to the write head needs to be weaker than the original signal, otherwise the system becomes saturated. One characteristic property of tape and plate systems is that the signal becomes weaker every time it is fed back through the system. This imitates physical reality, where the power of an audio signal gradually fades and the trebles are quickly absorbed as they are reflected by reflective surfaces (walls, partitions, arches, pillars, etc.).

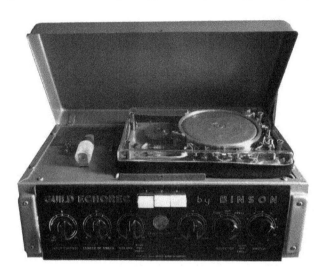

Figure 8.22. *The electronic tube echo chamber "Echorec" (1962) by Binson. You can see the plate surrounded by multiple heads, with a rubber drive wheel (in the center). In the middle of the controls, you can see a magic green eye. This controls the volume of the audio signal*

These systems were popularized by manufacturers such as Binson (Italy), Dynacord (Germany), Echoplex (USA), Fender (USA), Meazzi (Italy), Roland (Japan), Selmer TrueVoice (UK), WEM (UK) and many others, from the 1960s until today (Fulltone – USA).

8.1.4. *Reverb configuration*

Now that we have presented the theory, history, principles and technology of the various types of reverb that you can find on the market, we can finally begin to discuss their settings and parameters.

8.1.4.1. Main settings

The reverb time or RT60

– This is the duration for which the sound signal can still be perceived after the original signal has disappeared. Formally, it is defined as the time required for the signal to decrease by 60 dB (which explains the acronym RT60 – Reverb Time 60 dB) relative to the original signal (direct sound).

The predelay

– This is the time that elapses before the first reflections, known as the *early reflections*, are created in the audio signal. In physical contexts, this parameter depends on the physical dimensions of the room and any obstacles that it contains. The closer the nearest obstacle or surface, the shorter the predelay time. The higher the predelay time, the larger the size of the virtually simulated room. In a certain sense, this parameter "airs" out the signal by separating the dry sound from the wet reverb.

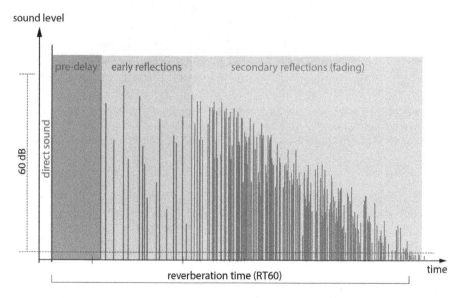

Figure 8.23. *Three main components of reverb. For a color version of this figure, see www.iste.co.uk/reveillac/soundeffects.zip*

The early reflections or initial level:

– These are the initial reflections, the responses from when the audio signal first encounters reflective obstacles. If the room or location is massive (like a valley or a tunnel), these reflections are echoes, but in most conventional venues (empty room, church, cathedral, etc.) they will be perceived as normal reverb. This parameter is directly related to the decay level of the reverb. If the first reflections are very weak, you are essentially simulating movement between the back and the center of the sound space from the perspective of the observer.

The decay (or decay time) or attenuation:

– This parameter plays a role in the *secondary reflections*, i.e. the reflections of reflections, creating a diffusion phenomenon among the sound waves. As the number of reflections increases, the intensity decreases until it vanishes completely as the waves are absorbed by the sound space. Decay is one of the fundamental parameters of reverb.

8.1.4.2. *Specific parameters*

Since audio processing is never more than an imitation of physical reality, convolution and algorithm-based reverb processors and plugins use an array of other parameters to describe the desired nature and texture of the effect more accurately.

We will see a few of these parameters in the following, but the list is far from exhaustive. Manufacturers and software publishers are constantly innovating and outdoing each other to offer ever more sophisticated, powerful, user-friendly and accurate products based on state-of-the-art scientific advancements.

The parameters presented in the following are listed in alphabetical order.

Crossover frequency

– This defines a threshold between high and low frequencies. This value of this parameter affects the other parameters.

Damping

– This characterizes the reflections in terms of the underlying materials that create them. For example, a hard surface will have a greater dynamic range and will be more favorable for high frequencies. A softer, gentler surface on the other hand will absorb these higher frequencies, creating a warmer sound. Adjusting the damping modifies the texture of the reverb, ranging from an artificial-sounding effect to a more natural one.

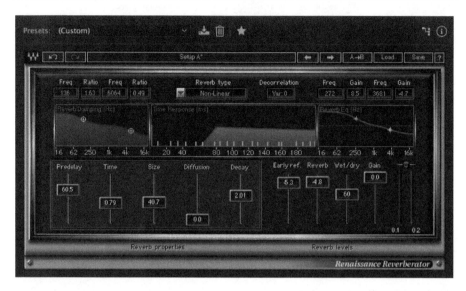

Figure 8.24. *Example configuration of a reverb plugin, the "Renaissance Reverberator" by Waves. This plugin includes most of the parameters listed in this section: damping, predelay, time, size, diffusion, decay, early reflections, mix (wet/dry), with various curves, equalization, reverb damping and the overall shape of the reverb (in the center). For a color version of this figure, see www.iste.co.uk/reveillac/soundeffects.zip*

Density

– This parameter increases or decreases the time interval between the first group of reflections and the next group. The nature of the sound signal is critical when configuring this parameter. Typically, drier sounds (very short attack) need higher densities and vice versa. Density can be viewed as the thickness of the reverb, and as such is a relatively similar concept to the diffusion.

Diffusion

– This parameter modifies the overall effect, and plays a key role in defining the rendering of the reverb. When the diffusion is increased, the early reflections are more tightly bunched together, which increases the mass of the sound signal. This makes echoes more difficult to distinguish individually. The best setting for this parameter therefore heavily depends on the nature of the sound. For pulse signals (percussion instruments), the diffusion typically needs to be high, whereas lower tends to be better for longer, sustained signals.

Time Effects 223

Figure 8.25. *Another example of reverb settings, this time the "LexRoom" by Lexicon. This plugin also includes many of the most common parameters. For a color version of this figure, see www.iste.co.uk/reveillac/soundeffects.zip*

Frequency decay (or *frequency attenuation/frequency level*)

– This parameter is often split into two or more distinct frequency bands (bass, mids, treble, etc.). In physically realistic sounds, higher frequencies are more strongly affected by the decay. However, there is no reason you cannot choose other configurations to create more unreal sounds. The frequency decay is proportional to the elapsed time. It play an important role in the overall sound of the effect.

Gate

– This defines a threshold for the signal level below which the reverb is inactive.

Gate decay

– This determines how long it takes for the reverb to fade once the signal passes below the threshold defined by the gate.

Mix

– This parameter determines the ratio of original signal (*dry*) to reverb (*wet*).

Figure 8.26. *A third example of the configuration of a plugin. This time, "VintageVerb" by Valhalla. For a color version of this figure, see www.iste.co.uk/reveillac/soundeffects.zip*

Preset (algorithm)

– The choice of reverb algorithm defines the IR. Each algorithm typically attempts to simulate a certain class of realistic acoustics, or in some cases virtual acoustics (acoustics that do not exist in reality). In digital reverb processors, pedals and plugins, the algorithms are usually saved as presets. Table 8.4 lists the most common types of reverb algorithm.

Type	Description	Representation
FLAT	The amplitude curve of the early reflections is flat. This simulates a small room. In fact, this approximately recreates the results of EMT-type plate reverb (see section 8.1.3.1).	
ROOM	The amplitude curve of the early reflections decreases according to a regular trend, more or less linearly. This simulates a relatively large room, like a small events hall or concert hall.	

HALL	The amplitude curve of the early reflections decreases very irregularly. This simulates a highly complex and vast propagation environment such as a cathedral, a very large hall, a railway station, an atrium, etc.	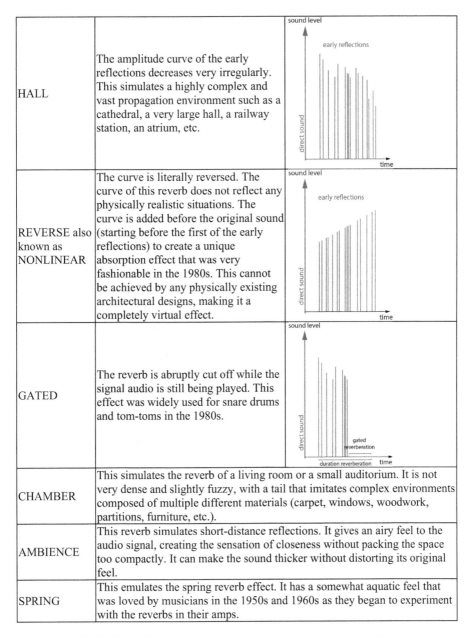
REVERSE also known as NONLINEAR	The curve is literally reversed. The curve of this reverb does not reflect any physically realistic situations. The curve is added before the original sound (starting before the first of the early reflections) to create a unique absorption effect that was very fashionable in the 1980s. This cannot be achieved by any physically existing architectural designs, making it a completely virtual effect.	
GATED	The reverb is abruptly cut off while the signal audio is still being played. This effect was widely used for snare drums and tom-toms in the 1980s.	
CHAMBER	This simulates the reverb of a living room or a small auditorium. It is not very dense and slightly fuzzy, with a tail that imitates complex environments composed of multiple different materials (carpet, windows, woodwork, partitions, furniture, etc.).	
AMBIENCE	This reverb simulates short-distance reflections. It gives an airy feel to the audio signal, creating the sensation of closeness without packing the space too compactly. It can make the sound thicker without distorting its original feel.	
SPRING	This emulates the spring reverb effect. It has a somewhat aquatic feel that was loved by musicians in the 1950s and 1960s as they began to experiment with the reverbs in their amps.	

Table 8.4. *The eight most common types of reverb algorithm*

Room size (or just *size*)

— This adjusts the reverb time to reflect the size of the virtual room that is being simulated. In larger rooms, the sound waves travel for longer before being reflected by an obstacle (wall or partition).

Spread

— This parameter widens or reduces the stereo image of the *reverb tail*[11]. When the spread is set to minimum, the sound is distributed monophonically. At maximum spread, the sound is distributed across a very wide stereo field.

Figure 8.27. *Another plugin by LEXICON, "Plate" reverb in LXP Native*

Width

— This increases or decreases the width of the stereo image.

REMARK.— As noted above, convolution reverb effects tend to be less flexible, with fewer configurable parameters. You can see an example of this in Figure 8.28.

---

11 The reverb tail (or "*diffuse reverb*") is the type of reverb obtained when there are so many reflections that they seem to blend together in time, creating a homogeneous sound with a diffuse and highly dense feel.

Figure 8.28. *The "Altiverb 7 XL" plugin by Audio Ease for convolution reverb. There are few parameters, but the most common parameters are still present: time, damping, predelay, mix, etc. For a color version of this figure, see www.iste.co.uk/reveillac/soundeffects.zip*

8.1.5. *Recording the IR and deconvolution*

As explained in section 8.1.3.4, there are two possible methods of determining the IR.

The "transients" method is difficult to implement. It can be very delicate to record a signal like a gunshot or a balloon bursting without distortion, since these sounds are very loud and have very strong transient components. Furthermore, this process yields little information about the lower (bass) and upper (treble) frequency bands, since they are not strongly represented, which limits the width of the usable range of reverb.

The "sine sweep" method is therefore preferable. This method covers the entirety of the audible frequency spectrum. Unlike the gunshot, we do not reuse the recorded sound signal directly. Instead, we first apply deconvolution to the recorded signal to extract information about the reflections of the space from it. This recorded signal is not a single impulse, but a continuous signal that sweeps over the entire spectrum.

We should stop for a second to clarify one important point. The recording process involves various pieces of equipment, including speakers, amplifiers, preamplifiers, microphones and analog-to-digital converters. The performance of each device influences the quality of the recorded sound, and so the quality of the IR necessarily depends on every piece of equipment involved in recording it.

8.1.5.1. *Configuring the sine sweep recording*

First, we need one sound source for each channel. For example, if we plan to apply the reverb to mono signals, we only need one speaker. For stereo signals, we need two speakers[12] and so on.

The recording can be implemented in several ways, but the simplest approach is to have one microphone for each source. This avoids the need for multiple recordings.

There are too many possible ways to position the sources and microphones for us to list them all. The ideal positioning depends on the desired results, the configuration, and location in which you are recording (size, materials, obstacles, etc.). Experience and experimentation will determine the rendering that you ultimately manage to achieve.

Here are some examples (the arrows denote the possible positions and orientations of the sources, in blue, and the microphones, in red):

– mono signal (Figure 8.29) – one speaker and one microphone;

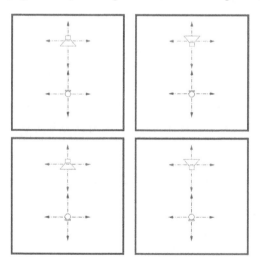

Figure 8.29. *Four example setups for recording a mono signal (there are many others): one speaker and one microphone facing each other; one speaker and one microphone facing away from each other; one speaker and one microphone facing in the same direction; one speaker and one microphone facing in the other direction. For a color version of this figure, see www.iste.co.uk/reveillac/soundeffects.zip*

---

12 If the IR is calculated using the "transients" method, no speakers are required. Instead, you use a balloon or a starter pistol as the sound source.

– stereo signal (Figure 8.30) – two speakers and two microphones;

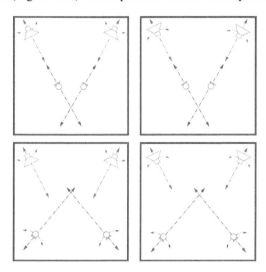

Figure 8.30. *Four example setups for recording a stereo signal (many others are possible): two speakers and two microphones facing each other (stage – audience); two speakers and two microphones facing away from each other; two speakers and two microphones facing in the same directions; two speakers and two microphones facing in the other directions. For a color version of this figure, see www.iste.co.uk/ reveillac/soundeffects.zip*

– five-channel surround sound (Figure 8.31) – five speakers and five microphones.

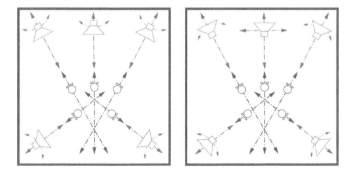

Figure 8.31. *Two examples setups for recording five-channel surround: five speakers and five microphones facing each other and five speakers and five microphones facing in the same directions. For a color version of this figure, see www.iste.co.uk/reveillac/soundeffects.zip*

Pointing the microphones toward the speakers creates a setup that imitates an orchestra or stage musician facing an audience. This is one of the most classic recording setups. When the speakers are oriented away from the microphones, there are no direct paths between them. This means that the recording will pick up many more reflections, and so the character of the location will be much more pronounced in the recording.

You can find download free "sine sweep" files from the following websites[13] for your own recordings:

– the Waves Ltd website (www.waves.com) under "Downloads" – "Free downloads": "IR convolution reverb sweep file for self-capture";

– the Audiocheck website (www.audiocheck.net), under "Test tones": "Full sine sweep";

– the Freesound website (www.freesound.org), under "Sounds": "Utility sounds: Sine Sweep 20 Hz to 20 kHz 10 seconds.wav";

– and many others that are easy to find with a search engine.

8.1.5.2. *Processing the IR*

Once you have completed the sine sweep recording, you will need a software program to perform deconvolution.

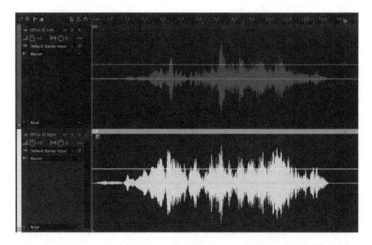

Figure 8.32. *The waveforms are two impulse response files from a stereo recording. The top and bottom are the left and right channels, respectively. For a color version of this figure, see www.iste.co.uk/reveillac/soundeffects.zip*

---

13 These websites were live at the time of writing.

For example, the following programs can do this:

– CATT GratisVolver – Microsoft Windows – freeware;
– Voxengo Deconvolver – Microsoft Windows – shareware;
– Christian Deconvolver – Microsoft Windows – freeware;
– Waves IR-1 – OS-X and Microsoft Windows;
– Audio Ease Altiverb – OS-X and Microsoft Windows.

8.1.5.3. *Example of deconvolution with Waves IR-1*

For our first demonstration of processing an audio file with the IR, we will use the multitrack audio editor Adobe Audition CC with the IR-1 plugin by Waves:

– after starting up Adobe Audition, open the file to which you want to add reverb;

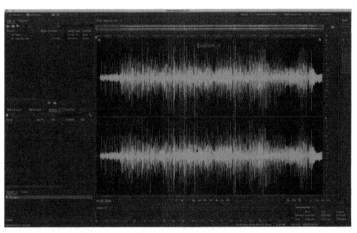

Figure 8.33. *Stereo audio file opened with Adobe Audition CC ready to add reverb. For a color version of this figure, see www.iste.co.uk/reveillac/soundeffects.zip*

– now, start up the Waves IR-1 plugin, and select the type of audio file (mono, stereo, live, etc.). This will open a settings window (Figure 8.34);

– in the example shown in the figures, the file is stereo, which means that two IR files are required, one for each of the left and right channels;

– next, create a new subfolder under "IR1Impulses V2" in the root folder "Waves", and copy your two IR files into it;

Figure 8.34. *IR-1 plugin window (in this example, the type is mono/stereo). For a color version of this figure, see www.iste.co.uk/reveillac/soundeffects.zip*

– click the LOAD button in the top right and select IMPORT SWEEP RESPONSE FROM FILE;

– a prompt will allow you to select the two sine sweep recording files;

Figure 8.35. *The prompt for selecting the sine sweep recordings. In the example, you can see the two files "SSL.wav" and "SSR.wav" in the newly created "IR Sine Sweep" folder. For a color version of this figure, see www.iste.co.uk/reveillac/soundeffects.zip*

– after a brief deconvolution calculation period, a new curve will appear in the center of the window of the IR-1, showing the IR of your reverb;

– you can now save your IR by clicking the SAVE button and selecting SAVE TO NEW FILE;

– enter a filename, and click SAVE (Figure 8.36);

Figure 8.36. *Saving the file (.xps) containing the preset for your new reverb effect (in this example, the file has been given the name "My Office IR.xps"). For a color version of this figure, see www.iste.co.uk/reveillac/soundeffects.zip*

– another dialog box will open, this time asking you to choose a name for the preset. The name you choose will be added to the list of options in IR-1. After clicking OK, your new preset will now be available at any time.

Figure 8.37. *Saving the new preset, here "Office reverb". For a color version of this figure, see www.iste.co.uk/reveillac/soundeffects.zip*

REMARK.– You might notice that a .wir file has been created in the folder containing your IRs. This is the deconvolution file generated by Waves IR-1.

If you restart IR-1, you will see your new preset as an option in the LOAD list.

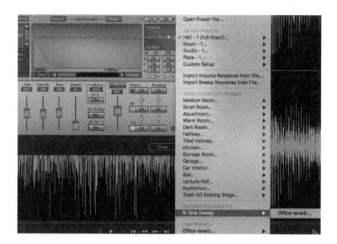

Figure 8.38. *The new preset "Office reverb" in the folder "IR Sine Sweep" that was created in an earlier step. For a color version of this figure, see www.iste.co.uk/reveillac/soundeffects.zip*

8.1.5.4. *All-in-one with Apple Logic Pro X*

The software program Apple Logic Pro X includes an "Impulse Response Utility". This application offers multitrack audio recording, deconvolution and can generate the IR using the reverb module "Space designer":

– start up Logic Pro X, create an EMPTY PROJECT, then choose to record an audio signal (Figure 8.39);

REMARK.– Choose the most appropriate digital audio input/output interface (M-Audio, MOTU, PreSonus, Apogee, FocusRite, etc.) that is available to you.

Figure 8.39. *Creating an audio project in Logic Pro X*

– display the mixing desk: press the X key or click VIEW then SHOW MIXER in the menu, or alternatively press the sixth icon from the left in the top ribbon;

– click AUDIO FX for input 1 on the mixing desk, select REVERB, SPACE DESIGNER, then MONO/STEREO;

– a dialog box will open for the convolution reverb plugin "Space designer". Scroll down the IR SAMPLE list and select OPEN IR UTILITY (Figure 8.40);

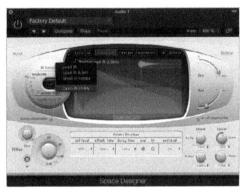

Figure 8.40. *Opening the "Impulse Response Utility"*

– this opens the "Impulse Response Utility";

– configure the number of tracks that you wish to record: mono, stereo, true stereo, pro-logic, etc., then click OK;

– now you can generate and record the audio signal of the sine sweep, which yields the IR;

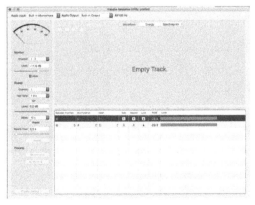

Figure 8.41. *The Impulse Response Utility*

REMARK.– The top toolbar of the utility allows you to configure the audio input and output devices. The sampling frequency displayed to the right of the audio output is for information only. You cannot configure it here, as it is determined by the choice of digital audio interface.

– connect the microphone(s) to your digital audio interface, one for mono or several for stereo or higher, and hook up a sound reproduction system to the outputs;

– arrange the equipment (microphones and speakers) around the room to capture the IR;

– in the SWEEP section, choose the CHANNEL, and select a TEST TONE of 1 kHz. Check the ON checkbox and adjust the LEVEL between –60 and 0 dB until the intensity is at the right level (no saturation on the level monitors, below the red region);

– check the intensity at other frequencies (100 Hz, 5 kHz, 10 kHz), adjusting the level if necessary;

– repeat for each channel;

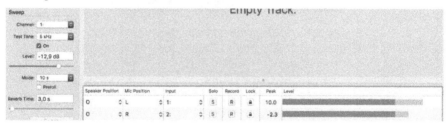

Figure 8.42. *Testing the level (–12.9 dB) at a frequency of 10 kHz on channel 1. The level monitors (green and yellow) of both channels are shown. For a color version of this figure, see www.iste.co.uk/reveillac/soundeffects.zip*

– uncheck the ON checkbox to end the test tone;

– click the R icon in the RECORD column of each channel. The background of the letter R will turn red;

– click the SWEEP button and wait for 10 s[14]. The sound signal will play, and you will see the waveform being plotted in the top window;

REMARK.– Make sure to turn off the monitor output to prevent feedback during recording. To do this, click MUTE.

---

14 Silence is of course required for this step, otherwise you risk distorting the recorded signal.

– after sound playback is complete, a dialog box will open, asking you to choose a filename for the IR. Note that the channels are automatically locked (green padlock in the LOCK column) in the table if you take additional recordings in order to keep the same settings for the sine sweep generator.

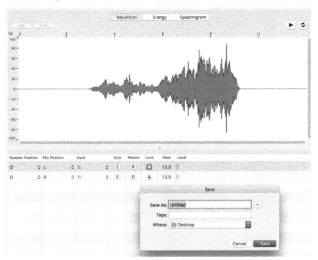

Figure 8.43. *The waveform of the recorded sine sweep and the dialog box for saving it. Track 1 is locked in the table. For a color version of this figure, see www.iste.co.uk/reveillac/soundeffects.zip*

– click on DECONVOLVE to begin deconvolution, which calculates the IR of your new reverb.

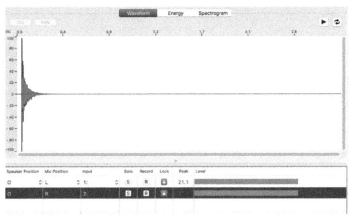

Figure 8.44. *The impulse response (IR) obtained by deconvolution*

To check whether your IR is good enough, you can use the listening tool, which allows you to apply convolution to test files and listen to the results.

– click on AUDITION IR to open the testing window;

– click on one of the four buttons SNARE, WOODBLOCK, MALE SHOUT or FEMALE SHOUT to test the rendering of your reverb;

– the MIX slider adjusts the balance between the test sound (source) and the sound after applying reverb. This equivalent to the "dry/wet" setting commonly found with reverb effects;

– the MONITOR slider adjusts the playback volume;

Figure 8.45. *Audio test window*

– to exit the audio test window, click DONE.

You can load one to five custom audio files by clicking the OPEN buttons. You can then listen to them by pressing PLAY.

You can save your IR by generating a .sdir file and a .pst file:

– click on CREATE SETTING..., choose a name for your file, then click OK;

– a dialog box will confirm that you have created the two files, and will display the path to their locations;

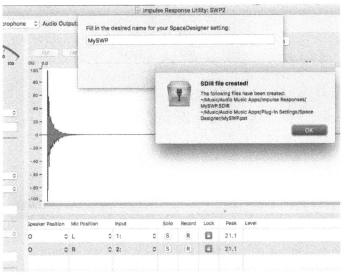

Figure 8.46. *Saving the IR, here with filename "MySWP"*

– click OK to confirm.

REMARK.– The "Impulse Response Utility" has various features that we have not discussed here, including IR editing, B-format surround encoding, the method of "transients" (using a "gunshot" instead of deconvolution), etc.

You can exit the "Impulse Response Utility" application by closing the window, clicking the IMPULSE RESPONSE UTILITY menu, then selecting QUIT IMPULSE RESPONSE UTILITY, or by pressing the keyboard shortcut CMD Q. A dialog box will ask whether you wish to save your project:

– click DON'T SAVE or SAVE accordingly.

Exiting the "Impulse Response Utility" returns you to the "Space designer" effects plugin. To use the newly generated reverb, we need to load the IR file:

– scroll down the IR SAMPLE list and select LOAD IR;

– in the dialog box that opens, select your .SDIR file, then click OPEN;

– your IR will be displayed, and you can now use it however you wish by configuring it and applying it to one or multiple audio tracks in Logic Pro X.

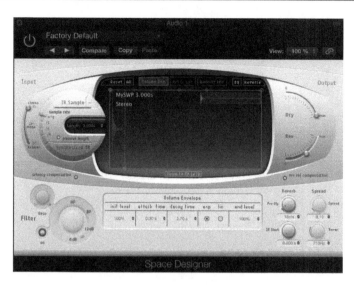

Figure 8.47. *Our newly created reverb, given the name "MySWP" in this example, viewed in "Space designer" within Apple Logic Pro X*

8.1.6. *Studio mixing and reverb*

Managing reverb when mixing a recording is a complicated task that requires precision, experience and finesse. The primary objective is to give each instrument its own position within the sound space.

In the following, we give a few guidelines that should help you to understand the basic ideas of reverb management.

First, we shall consider the effect of the predelay parameter, which allows the position of an instrument to be defined.

There is a simple rule for this:

– shorter predelay times (between 10 and 20 ms) make the instrument seem further away from the observer. This places the instrument at the back of the mix;

– longer predelay times (>20 ms) make the instrument seem closer to the observer. This places the instrument at the front of the mix.

Do not set the predelay too high. If you do, it turns into a delay, which creates a noticeable duplicate signal, especially for very short sounds (percussions). The predelay is expressed in milliseconds.

Another important parameter is the diffusion. In simplified terms, the diffusion controls the mass of the reverb. At its lowest setting, there are few reflections. This creates the impression that the sound signal is being played in a highly absorbent room. Choosing the highest setting creates the impression of an extremely reflective room. The diffusion is often expressed as a value between 0 and 100.

The *damping* also gives the reverb a very particular color. Note that this setting is not available on every model of reverb. The damping allows you to configure the reverb time on one or multiple frequency bands.

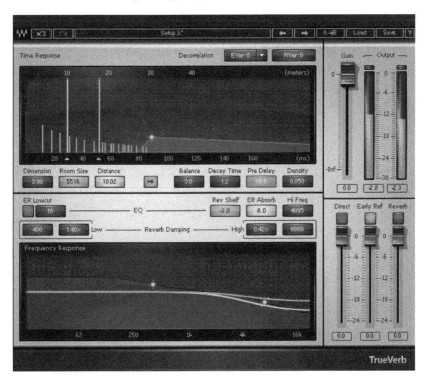

Figure 8.48. *The "TrueVerb" plugin by Waves. The bottom graph shows the frequency response curve of the damping. In this example, the damping is configured to increase the reverb time by 1.40× (140%) at 400 Hz (bass frequencies) and reduce the reverb time by 0.42× (42%) at 6,800 Hz (treble frequencies). For a color version of this figure, see www.iste.co.uk/reveillac/soundeffects.zip*

The *size* (or *room size*) is a very subjective parameter. The person managing the mix needs to rely on their own judgment to create the right atmosphere, depending on whether they want a large sound space or a narrower one.

In mixes that already have reverb on one or more tracks, you will need to implement any additional reverb as a postfader effect on the corresponding auxiliary track.

Finally, Table 8.5 gives a few examples of reverb configurations, drawing from my own personal preferences and experience (sound recordings taken in my studio: rectangular surface area: 40 m^2 – height: 2.90 m – floor: wood + carpet – walls: concrete + acoustic panels – ceiling: acoustic tiles).

Remember that the primary purpose of reverb is to define regions within the sound space and improve the presence and color of the sound source. Use it sparingly – too much reverb degrades the clarity and precision of the mix.

Be careful when adding reverb to sounds that already have reverb. It can be done, but you need to dose it carefully, or you might create resonance, phase variations, saturation and many other undesirable side effects.

Source	Style	Type	Predelay (ms)	Diffusion
Electric guitar	Solo rock	Hall	60–80	max
	Rhythmic rock	Room	40–60	max
	Bass guitar rock	Room	30–40	max
Electric guitar	Solo blues	Flat	60–70	max
	Rhythmic blues	Flat	40–60	max
	Bass guitar blues	Room	25–35	max
Acoustic guitar	Solo blues	Flat	40–50	max
Drums	Snare	Room	0–10	max
	Bass drum	Ambiance	0–10	3/4-max
	Cymbal	Chamber	0–10	3/4-max
	Tom-toms	Room	0–10	max
Vocals	Rock	Flat-hall	30–50	1/2
	Blues	Flat-chamber	20–40	max

Table 8.5. *Example configurations for a selection of sources and styles*

It is advisable to use as little reverb as possible when recording, to the extent that your recording location allows you to. It is much easier to add reverb later than remove it.

8.2. Delay

In section 8.1.3.5.2, we talked about echo chambers based on tapes or plates. These echo chambers are sometimes used for reverb by using high scrolling speeds and positioning the write and read head(s) very closely together. But these chambers are in fact perfect for creating delays; it is what they do best.

Delay recreates the everyday phenomenon of echoes. An echo is when a sound repeats several times, fading over time. If you could write an echo in words, it might look something like this:

– you shout ECHO...

– ...and a few tenths of a second later, your surroundings answer you back:

– ECHO... ECHO... ECHO... echo... ech... ec...

This repetition is caused by the sound returning to you after being reflected somewhere far away. The sound is quieter when it comes back because part of its energy was absorbed when it collided with the reflecting material. Once the reflected sound reaches you, the process starts all over again. The sound is reflected back and forth several times until the volume fades completely.

The difference between reverb and an echo or delay is that you can distinguish each repetition with echoes. With reverb, the repetitions merge together and are indistinguishable.

REMARK.– There is a technical distinction between a delay and an echo. A delay can have infinitely many repetitions, whereas an echo cannot. However, in everyday language, the terms "echo" and "delay" are often used interchangeably.

8.2.1. *History*

The first echo effect to be used in a recording is often misattributed to Sam Phillips, the CEO of Sun Records. In fact, Les Paul and Mary Ford had used echoes before that in 1951 for their track "How High The Moon".

In 1954, Sam Phillips used two tape recorders (Ampex 350), modifying one of them to allow that the write and read heads to be activated simultaneously. Doing this led him to invent a form of tape delay known as "*slapback*".

Slapback delay works as follows:

– the write head of the first tape recorder records the local signal picked up by the microphone;

– a read head is placed slightly behind the write head on the tape. The read head plays the newly recorded signal with a slight offset (a few centimeters). This creates a delay or echo (the delay time depends on the scrolling speed of the tape – at the time, the speed was most likely 7.5 inches per second, or around 19 cm/s);

Figure 8.49. *The famous Ampex 350 recorder*

– the mixing desk sends these two signals (the original signal from the microphone and the delayed signal from the first tape recorder) to the second tape recorder, which records them together. Of course, this delay effect was only capable of creating a single repetition.

Over the course of the 1970s, many musicians began to experiment with tape delays. Some of their solutions used highly unusual techniques (Frank Zappa, Pink Floyd, Robert Fripp, etc.).

8.2.2. *Types of delay*

There are four broad categories of delay:

– tape or plate delay;

– analog delay;

– digital delay;

– software delay (plugins).

We talked about tape and plate delays earlier.

The next category, analog delays, was invented in the 1970s because of the progress that had been made in musical electronics. The delay time of these effects was limited, usually 500 ms at most.

Some configurable settings were already available, e.g. the mix, which defines the ratio of direct and processed sound (*wet/dry*), and the *feedback*, which specifies the number of repetitions.

One major problem with these delay effects is that they introduce a certain color in the sound, often combined with slight distortion.

They are often integrated into effects pedals.

Figure 8.50. *The famous "Memory Man" pedal by Electro Harmonix, one of the first ever analog delay pedals (delay time: max 300 ms)*

The first digital delays began to appear in the early 1980s. Many of them are still available today from various manufacturers. They offer great sound quality and dynamics in the form of effects pedals or studio racks. They have much more sophisticated settings, with delay times ranging up to several seconds.

Figure 8.51. *The digital delay DM1000 by Ibanez, released in the early 1980s. It has a maximum delay of 900 ms*

These settings can include:

– a *sample* or *hold* function, which saves repetitions in memory;

– a filter to attenuate specific frequency bands;

– a low-frequency oscillator (LFO) with configurable depth and modulation speed to modulate the repetitions;

– a "*tap*" or "*tap-tempo*" function to specify the time between repetitions (allowing the delay to be synchronized with the beat) via a tap button;

– and many more.

Figure 8.52. *A modern digital delay pedal, the DD-500 by Boss. This pedal has a very wide range of possibilities. It can even emulate digital delays from the 1980s*

Today, digital delay effects have become extremely sophisticated, with delay times of up to 10 s. They often have integrated looping features, and can store patches with factory or user presets. Some include USB and/or MIDI ports in addition to the usual audio inputs/outputs.

Most multieffects racks available on the market include some form of delay, with varying degrees of sophistication.

Table 8.6 lists some widely used examples of analog and digital delays. The list is far from exhaustive, as the range of available models is simply too vast. Delays can be divided into two broad categories: studio delays in the form of racks[15] and delays in the form of pedals designed for musicians, most notably for guitarists.

Category	Manufacturer	Model	Year	Main features
Rack	Ibanez	DM1000	1983	Digital – Delay time: 1.75–900 ms
	Ibanez	DM2000	1983	Digital – Delay time: 0–1,023 ms
	Korg	SDD3000	1982	Digital – Delay time: 0–1,023 ms
	Lexicon	PCM-42	1981-82	Digital – Delay time: 400–4,800 ms
	Lexicon	MPX-1	1990s	Digital multieffects – Delay time: 0–2,000 ms
	Lexicon	MPX-550	2003	Digital multieffects – Delay time: 0–5.5 ms for mono, or 2.7 ms for stereo
	Lexicon	PCM-92	2009	Digital – Reverb + delay + modulation effects
	Lexicon	Prime Time II	1979-80 Reissued in 1995	Digital – integrated LFO – Delay time: 1.92 s max.–7.68 s max. with additional memory
	Mu-tron	1173	1980s	Digital – Delay time: 2.5–160 ms
	Roland	SDE-3000	1983	Digital
Pedal	Boss	DM-1	1978	Analog – Delay time: 500 ms max.
	Boss	DM-2	1981	Analog – Delay time: 20–300 ms
	Boss	DSD-2	1985	Digital – Includes a sampler function
	Boss	DD-3	1983	Digital – Delay time: 20–300 ms
	Boss	RV-2	1987	Digital – Includes a panning function ("ping-pong")
	Boss	RV-3	1994	Digital – Delay time: 2 ms max.
	Boss	DD-6	2002	Digital – Delay time: 5.2 ms max.
	Boss	DD-7	2008	Digital – Delay time: 6.4 s max.
	Boss	DD-500	2015	Digital – Various highly advanced functions
	Electro	Deluxe	2010	Analog – Delay time: 30–550 ms –

---

15 This list does not include tape or plate delays, which are presented in section 8.1.3.5.2.

Electro Harmonix	Memory Boy		Includes a tap function
Electro Harmonix	Memory Man	1976-1978 – Multiple reissues in 1990 and after	Analog – Includes chorus and vibrato – Very colorful sound
Electro Harmonix	Memory Man XO	2009	Analog – Delay time: 700 ms max. – Includes chorus and vibrato
Electro Harmonix	Memory Man with Hazarai	2008	Analog – Delay time: 3 s max. – Loop, reverse, multitap functions
Ibanez	AD9	1982	Analog – Delay time: 20–300 ms
Korg	SDD3000 pedal	2014	Digital – Emulates the analog SDD3000 and many others
Moog	MF-104	2000	Analog – Delay time: 40–800 ms – Integrated LFO
MXR	M169 Carbon copy	2008	Analog – Delay time: 20–600 ms
TC Electronic	Flashback	2014	Digital – Delay time: 20–7,000 ms – Many features
Vox	Time machine	2008	Digital emulation of analog delay – Delay time: 0–1 s

Table 8.6. *Examples of delay racks and pedals*

Next, let us talk about plugins. Just like reverb, there are countless software-based delay effects. Many of them are virtual recreations of hardware delays available on the market.

Figure 8.53. *The plugins PSP 42 and PSP 84 by AudioWare, inspired by the famous PCM 42 delay by Lexicon*

There are many things that you can do with these plugins. Most of them offer plenty of settings for configuring the standard delay parameters. For example:

– a *panning* (or "*ping-pong*") effect;

- an LFO;
- multiple filters;
- an equalizer;
- chorus, phaser, distortion;
- a tap-tempo function (for synchronizing the delay to the beat);
- a loop function;
- etc.

Figure 8.54. *The Manny Marroquin plugin by Waves, a very sophisticated delay effect that includes reverb, distortion, phaser and a doubler*

Table 8.7 lists a few examples of popular delay plugins.

Publisher	Name	Operating system and format
Arne Scheffler	Multi Delay	OSX: VST (freeware)
Audio Damage	DubStation	Windows: VST OSX: AU/VST
AudioWare	PSP 85	Windows: VST; RTAS; AAX OSX: VST; RTAS; AAX; AU
Avid	Line6 Echo Farm	Windows – OSX: TDM
Big Tick	Hexaline	Windows: VST (freeware)
D16	Sigmund	Windows: VST; AAX OSX: VST; AAX; AU
FabFilter	Timeless 2	Windows – OSX: VST 2/3
Kjaerhus Audio	Classic Delay	Windows: VST (freeware)
MaxProject	T.Rex Delay	Windows: VST (freeware)
Ohm Force	OhmBoyz	Windows: VST; VST3; RTAS; AAX OSX: VST; RTAS; AAX
Rob Papen	RP Delay	Windows: VST; AAX OSX: VST; AAX; AU
Smartelectronix	Analog Delay	Windows: VST (freeware) OSX: VST (freeware)
SoundToys	EchoBoy	Windows: VST; AAX OSX: VST; AAX; AU
Universal Audio	Roland RE-201 Space Echo	Windows: RTAS; VST; AAX OSX: RTAS; VST; AAX; UA
Waves	H-Delay	Windows: VST; VST3; AAX; RTAS OSX: VST; VST3; AAX; RTAS; AU
Waves	Manny Marroquin	Windows: VST; VST3; AAX; RTAS OSX: VST; VST3; AAX; RTAS; AU
Waves	SuperTap	Windows: VST; VST3; AAX; RTAS OSX: VST; VST3; AAX; RTAS; AU

Table 8.7. *Examples of delay plugins*

8.2.3. *Tips for using delays in the studio*

8.2.3.1. *Slapback*

We mentioned this effect earlier in section 8.2.1. To generate slapback delay, select a delay time between 50 and 120 ms, and set the feedback to zero so that there is one single repetition very close to the source signal.

Additionally, we need to reduce the signal bandwidth acted upon by the effect, otherwise the repeated signal will sound exactly like the original, creating an extremely unpleasant duplication effect.

To restrict the bandwidth, equalize the top and bottom of the spectrum by setting up appropriate filters.

Slapback delay is usually reserved for electric guitars and vocals. It endows them with a vintage quality.

8.2.3.2. *Doubling*

Doubling is a way of making the sound feel bigger.

Choose a very short delay time between 5 and 50 ms, set the feedback to 0 and adjust the dry/wet mix until you can only just hear the effect.

Above this limit, increasing the presence of the effect (making it *wetter*) will make the sound seem less natural. Conversely, decreasing the presence (making it *drier*) will make your sound seem more natural and occupy more space.

Beware of unpleasant filtering effects in the mix if you set the delay time too low.

Similar to slapback delay, you will typically need to equalize the delayed portion of the signal.

To make your sound seem even bigger, you can use stereo delay with slightly different delay times on each channel, or even apply a slight modulation.

Doubling is especially convincing with analog delays, as this trick was originally designed for analog systems.

8.2.3.3. Pan-delay

Also known as "ping-pong delay". With this delay, the reverb swings back and forth between left and right. If the delay time is very short, less than 70 ms, the effect becomes diffuse, creating a reverberating space.

Setting the delay time higher, above 120 ms, gives an artificial quality to the sound.

Figure 8.55. *The H-Delay plugin by Waves with filters, LFO, and a ping-pong delay function (top center)*

Pan-delay is included in many modern pedals and plugins.

8.2.3.4. Delay and tempo

If you synchronize the delay time with the tempo of the sound, you can add depth to the mix without the delay itself being noticeable. Many sound engineers use this method to stylize their sound environments, endowing them with a very natural ambient reverb that smoothly blends into the overall structure of the recording in the final mix.

Most reverbs have tap-tempo features. If your reverb does not include this feature, there are also dedicated apps (including ones that you can run on your smartphone) to find the right delay time for a given musical passage. Here is a list of examples[16]:

- Tap/Tempo by Benjamin Jaeger for iOS;
- BPM Tap Tempo by Audiolog for iOS (free);
- BPM Tap by Erik Byström for Android (free);
- Tap Tempo for Windows (free);
- Delay Genie by Bobby Owinski for iOS (free);
- Tap BPM by DrumBot for Windows and OSX (free);
- Tap Tempo and Metronome for Android (free).

8.2.3.5. Delay/echo with an analog multitrack tape recorder

This section will explain how to use a multitrack tape recorder to create a delay or echo effect.

Why use such outdated equipment today? I know for a fact that many audio enthusiasts and recording professionals still have tape recorders stashed away somewhere in a closet or a corner. This will give you the opportunity to dust them off and use them to create a vintage warmth and color that is currently very trendy.

You will need a multitrack tape recorder (2, 4, 8, 16 or more tracks) with separate read and write heads, e.g. a Revox A77 or B77.

Figure 8.56. *The stereo Revox A77 MkIV recorder, manufactured from 1974 to 1977*

---

16 These apps are still available at the time of writing. You may need to check if they are still accessible.

The idea is as follows. During recording, the signal is picked up by the read head and sent to the write head, which re-records the signal, delayed by a time interval determined by the distance traveled by the tape between the two heads.

This time interval and hence the delay time depend on the speed of the tape. For example, for a Revox A77:

– 1 ⅞ inch/s (4.75 cm/s): 700 ms;

– 3 ¾ inch/s (9.5 cm/s): 350 ms;

– 7 ½ inch/s (19 cm/s): 175 ms;

– 15 inch/s (38 cm/s): 87 ms.

Some models of recorder have a speed regulator. If yours does, you can use it to vary the delay/echo linearly over a certain range of speeds.

Figure 8.57. *A speed regulator made by Revox for the B77 model recorder*

In the following, we will describe the procedure for a Revox A77 or B77. They can easily be adapted to other models of equipment;

– connect the source to input II (mic input on the front or aux on the back);

– select this source by switching the channel dial II to MIC LO, MIC HI or AUX;

– the LEVEL II dial controls the level of the recorded sound signal. This should be set as high as possible, but saturation and excessive modulation should be avoided, as they will combine with the delayed signal;

– turn the dial for input I to I > II in order to send track I to track II;

– press REC CHI to confirm the recording channel, leaving REC CHII pushed out;

Figure 8.58. *The various controls of the Revox A77. From left to right on the bottom row: volume and mono/stereo dial, balance and MONITOR dial, recording track 1 and the dial for selecting its source, and recording track 2 with the dial for selecting its source. In the top right, you can also see the REC CHI and REC CHII recording buttons on either side of the two VU level meters.*

– the intensity of the delay effect is controlled by the LEVEL I dial. By turning this dial, you can pass from a short and weak delay to a long and powerful delay, and internal feedback;

– set the MONITOR switch to NAB and the mono/stereo dial to channel I;

– to begin recording, you can now simply press PLAY and REC together.

REMARK.– If you decide to record from channel I, you need to choose the opposite settings for the input dial (II > I). In this case, the source is controlled by the LEVEL I button, and the delay is controlled by the LEVEL II button. You also need to reverse the states of the recording buttons, and the dial specifying the mode needs to be set to II.

8.3. Conclusion

Reverb plays an extremely important role in sound, and our relationship with it is rather special. We constantly encounter reverb in every sound space of our everyday lives. Our brains analyze the tiniest of details contained in this reverb in order to deduce information about our surroundings.

Artificial reverb can help us build sounds that are more relevant, more catchy, warmer and easier to understand, but can also render sounds incoherent, artificial or inaudible.

Learning to master reverb is a difficult and complex process. There are many settings to manage, and it is crucial to dose every one of them accurately. The entire structure of a recording mix relies on the foundation provided by its reverb. Reverb is a fundamental component and significantly affects the final rendering.

We encounter delays and echoes less frequently in reality, which may explain why they are so very delicate to manipulate.

You will need to rely on your own personal judgment and ear to decide whether a given delay or reverb sounds good or bad. Determining precisely the right amount of reverb and delay will add presence to your sound and make it feel natural, as if you are standing in the room, the audience, at the concert, or even directly among the musicians on stage, depending on the rendering that you are trying to achieve.

Listen closely, close your eyes and imagine the scene. Your brain needs to be able to retrace the balance of the sound and the path that it takes. The path must be clear and easy to understand. This clarity is key in determining the flavor, character and charm of your mix.

9

Unclassifiables

This chapter attempts to group together the effects that do not fit into any of the six other categories, for example, because they act upon more than one of the fundamental characteristics of the audio signal: frequency, modulation, amplitude, dynamics, phase, tempo, etc.

This chapter also discusses specialized processing techniques used for tasks such as audio restoration, synchronization, sample editing and other activities that you might encounter while managing sound effects.

9.1. Combined effects

In this section, we will discuss three combined effects, i.e. effects obtained by combining several different techniques from the previous chapters. These three combined effects are called *fuzzwha*, *octafuzz* and *shimmer*.

As discussed before, we will begin by presenting the theoretical principles on which these effects are based, and then we will list a few hardware and software models.

9.1.1. *Fuzzwha*

This effect is a combination of wah-wah (see section 4.3) and fuzz (see section 7.6.1). It was very popular in music from the 1970s.

Most fuzzwha pedals allow you to select just wah-wah or just fuzz, but can of course also mix both effects. Fuzzwha is very rich in harmonics, with a somewhat distorted sound.

Below are a few examples of fuzzwha pedals. There are many others as follows:

- Fender: "Fuzz-Wah Pedal Reissue" model;
- Colorsound: "Wah-Fuzz" model (vintage);
- Aria Diamond: "Wah-Fuzz" model (vintage);
- Morley: "Power Wah Fuzz" model (vintage);
- Boss: "PW-10" model;
- Mooer: "FuWah" model;
- Morley: "M2 Cliff Burton Fuzz Wah" model;
- Electro Harmonix: "Cock Fight" model;
- Vox: "Stereo Fuzz-Wah Model 9-3700" model.

Figure 9.1. *Three fuzzwha pedals: (from left to right) "Fuzzwha" by Colorsound, "M2 Cliff Burton" by Morley and "FuWah" by Mooer*

9.1.2. Octafuzz

Also known as "*Octavia*", this is a fuzz pedal (see section 7.6.1) that also adds an extra harmonic one octave higher or lower (see section 6.2.1).

There are only a few models of this pedal. Examples include:

- MXR: "SF01 Slash Octave Fuzz" model;
- Catalinbread: "Octapussy" model;
- Fulltone: "Octafuzz" model;

- NIG Music: "Octa Fuzz POC" model;
- Joyo: "JF-12 Voodoo Octave" model;
- Fulltone: "Ultimate Octave" model;
- Home Brew Electronics: "UFO (Ultimate Fuzz Octave)" model.

Figure 9.2. *Three octafuzz pedals (from left to right): "Octafuzz" by Fulltone, "Octapussy" by Catalinbread and "SF01 Slash Octave Fuzz" by MXR*

9.1.3. *Shimmer*

This effect combines reverb (see Chapter 8) with an octaver (see section 6.2.4) to create a sound similar to synth pads.

It is very difficult to describe this effect in words. Its sound varies wildly from instrument to instrument and depending on musical technique (see the web links at the end of the book). The word "shimmer" comes fairly close to capturing its essence.

This effect is not usually created with a single device. Many guitarists create shimmer by coupling together multiple effects pedals or in some cases by combining effects pedals and software programs such as "IK Multimedia Amplitube" and "Native Instruments Guitar Rig". Some multieffects audio racks include shimmer directly as a preset.

9.1.3.1. *Creating shimmer*

Here are a few examples of combinations that you can use to construct the famous shimmer effect:

– Solution 1: Delay pedal (Boss DD7, MXR M169 Carbon Copy, Boss DD20, Electro Harmonix Canyon, etc.) coupled to a pitch shifter with a second delay effect added using AmpliTube (solution first proposed by Guitar Douche).

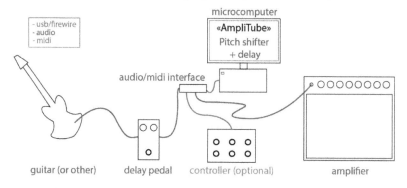

Figure 9.3. *First solution for creating shimmer. For a color version of this figure, see www.iste.co.uk/reveillac/soundeffects.zip*

– Solution 2: This solution was developed by Bill Rupert. It uses multiple Electro Harmonix pedals coupled together. The final sound rendering is unbelievable.

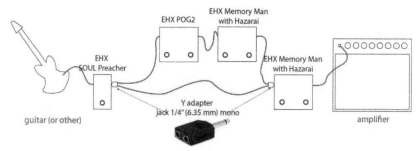

Figure 9.4. *The famous "Crystal Shimmer" by Bill Ruppert, using only pedals manufactured by Electro Harmonix. For a color version of this figure, see www.iste.co.uk/reveillac/soundeffects.zip*

– Solution 3: The third solution was proposed by Justin Gomez. It uses the "Guitar Rig" amp simulator in Native Instruments. The following modules are chained together in Guitar Rig: Volume–Compressor–Octaver–Pitch Pedal–Delay Man–Octaverb (see Figure 9.5).

Figure 9.5. *Shimmer created with "Guitar Rig". For a color version of this figure, see www.iste.co.uk/reveillac/soundeffects.zip*

9.1.3.2. *Examples of dedicated pedals*

Below is a non-exhaustive list of dedicated shimmer pedals and other pedals with a shimmer mode.

Shimmer pedals are much more difficult to compare than pedals for other effects, such as wah-wah, distortion, reverb and so on, since every version of shimmer produces a very different sound rendering. The only way to truly compare any two pedals is to try them both:

– Neunaber Technology, "Seraphim Mono Shimmer" model;

– Neunaber Technology, "Wet Stereo Reverb" model;

– Strymon, "blueSky Reverberator" model;

– Eventide, "Space Reverb" model;

– Boss, "RV6" model;

– MAK Crazy Sound Technology, "Space Reverb" model;

– Walrus Audio, "Descent" model;

– Mugig, "Reverburg" model.

Figure 9.6. *Four pedals with shimmer effects (from left to right): "Reverburg" by Mugig, "blueSky" by Strymon, "Space Reverb" by MAK Crazy Sound Technology and "Seraphim Mono Shimmer" by Neunaber Technology*

9.2. Tremolo

The tremolo effect is a variation in the amplitude of the sound created by means of an electromechanical or electronic system, as if you were manually turning the volume button up and down according to a certain rhythm.

Tremolo is often confused with the vibrato effect, but vibrato is in fact completely different (see section 6.1).

9.2.1. *History*

Tremolo is very old. Mechanical systems for producing this effect were probably invented in the 16th Century. One of the first known examples was a pipe organ in the San Martino Maggiore church in Bologna, Italy, dating from 1555. The way that the system was designed (by varying the air pressure in the pipes) created both tremolo and vibrato simultaneously, since the pressure affected both the volume and the pitch.

This may explain why there is still so much confusion between the two effects.

Over the next few centuries, various inventors filed patents for devices for tremolo only, vibrato only or both together, but their documents often failed to discriminate between the terms vibrato and tremolo, perpetuating the misuse of terminology.

Much more recently, in 1941, the company DeArmond[1] is thought to have invented the first ever electromechanical tremolo. Their device consisted of a small

---

[1] Named after Harry DeArmond, 1906–1999, American musician and inventor. In the mid-1930s, he invented the first ever commercial guitar microphone.

motor that worked by shaking a cylindrical container filled with an electrolytic liquid. Inside this (partly metal) cylinder, there was an electrode that intermittently became conductive whenever it was splashed by the liquid. To increase the agitation speed of the liquid, the device had a button for adjusting the position of a wheel along a conical axis, which changed the agitation speed of the cam to which the cylinder was fixed.

Figure 9.7. *The DeArmond tremolo (model 601) manufactured by ROWE Industries from 1946 onwards. On the right, part of the internal mechanism, showing the conical axis of the motor, the wheel and the cylindrical container filled with electrolytic liquid*

The first tremolos manufactured by DeArmond were designed to be mounted under the keyboard of Storytone[2] electric pianos, which were unveiled in 1939 at the New York World Fair a few years before the mechanism was marketed to guitarists.

Figure 9.8. *The Storytone piano*

---

2 Storytone pianos were the fruit of a partnership between Story and Clark and RCA. The instrument was designed by John Vassos, a renowned American industrial designer.

In the 1950s, many amplifiers began to offer integrated tremolo features. Tremolo was marketed as a selling point by brands such as Fender, Gibson and other manufacturers.

The "Tremolux" (1955) and "Vibrolux" (1956) models by Fender and the "Falcon" model by Gibson were among the first amplifiers to include tremolo effects.

Figure 9.9. *The "Tremolux" amp by Fender and the "Falcon" by Gibson*

Over time, as electronic circuits continued to improve, the quality of tremolos also steadily increased, creating better sound renderings.

9.2.2. *Examples of tremolos*

The tremolo effect is most popular with guitarists, for whom a wide range of tremolo pedals have been designed. It is also offered by most multieffects racks (see Appendix 4) and plugins. Tremolo effects often include a vibrato setting.

Figure 9.10. *Three tremolo pedals (from left to right): the "Supa-Trem2" model by Fulltone, the "TR-2" model by Boss and the "Super Pulsar" model by Electro Harmonix*

Table 9.1 lists a few examples of tremolo pedals and plugins.

Type	Manufacturer or publisher	Model	Remarks
Pedal	Boss	TR-2 Tremolo	Tremolo (configurable waveform)
	ColorSound	Tremolo 1975	Tremolo (vintage sound)
	Digitech	Snake Charmer Tremolo	Tremolo (three waveforms)
	Dunlop	TS-1 Tremolo	Tremolo (configurable waveform)
	Electro Harmonix	Super Pulsar	Stereo tremolo – 8 presets – 9 rhythm shapes – TAP function*
	Fulltone	Supa-Trem2	Tremolo – TAP function* – 3 waveforms – Stereo
	Hiwatt	Tube Tremolo	Tremolo (vintage sound)
	Ibanez	TL5 Tremolo	Tremolo
	Mooer	Varimolo	Tremolo – 3 modes
	Mooer	Spark Tremolo	Optical tremolo – 2 modes
	Line 6	Tap Tremolo	Tremolo – TAP function* – 3 modes
	Lovepedal	Tremelo	Tremolo – TAP function* – 3 modes
	Voodoo Lab	Tremolo	Tremolo (vintage sound)
Software	Adam Monroe Music	Tremolo	Freeware Windows: VXT; AAX OSX: VST; AAX; AU
	Audio Damage	PulseModulator	Freeware – Tremolo/distortion Windows: VST OSX: VST; AU
	MeldaProduction	MMultiBandTremolo	Windows: VST; VST3; AAX OSX: VST; VST3; AU; AAX
	Nomad Factory	Free Tremolo	Freeware – Part of Nomad Factory's free package
	Rekliner Records	TremoLlama	Freeware OSX: VST
	Slim Slow Slider	SimpleTremolo	Freeware Windows: VST
	SoundToys	Tremolator	Emulator for the tremolo on the Fender Vibrolux amp and the Wurlitzer electric piano Windows: VST; AAX OSX: VST; AU; AAX

*The TAP function allows you to synchronize the tremolo by tapping a rhythm (tempo) with your foot.

Table 9.1. *Examples of tremolo pedals and plugins*

9.3. Sound restoration tools

In this section, we will say a few words about sound restoration tools. Strictly speaking, these tools are not sound effects – instead, the goal is to remove flaws from an audio signal. Examples of such flaws include crackling, hissing, distortion caused by excessively high recording levels, undesirable noises on old vinyls or 78-rpm records caused by scratches and high levels of background noise on old magnetic tapes.

The introduction of software components (plugins) has made it much easier to process low-quality audio recordings in order to coax a little more life and energy out of them.

The following tools are essential for good sound restoration:

– *declicker*;

– *decrackler*;

– *denoiser*;

– *declipper*;

– *debuzzer*.

REMARK.– The functionality of some these tools overlaps, even though they are based on slightly different processes. For example, to eliminate a 50-Hz hum, you can use either a denoiser or a debuzzer interchangeably. Audio restoration terminology is often used relatively imprecisely. For example, some publishers and manufacturers use the term *declicker* to describe what others might call a *descratcher*.

9.3.1. Declickers

The purpose of this tool is to detect *clicks* and *pops* in the original signal, then eliminate them by synthesizing a replacement signal.

Clicks (snaps) and pops are very short, impulse-based noises with high amplitudes. They are often found on vinyl records, created whenever the microphone is bumped, as well as in degraded digital recordings, or caused by short interference on a wireless microphone. Clicks are rarely longer than 10 ms.

Figure 9.11. *The results of applying a declicker: the clicks in the shaded region are detected, eliminated and replaced by a synthesized signal (in green). For a color version of this figure, see www.iste.co.uk/reveillac/soundeffects.zip*

9.3.2. *Decracklers*

Much like declickers, decracklers identify crackling (static noise), and then mix it with the signals immediately before and after the crackling to significantly attenuate any undesirable noise.

Crackling has lower amplitude than clicks and lasts for longer. Conceptually, it is more of an irregularity in the continuous flow of the audio signal – a component that does not fit with the surrounding audio and music.

9.3.3. *Denoisers*

Denoisers are used to eliminate background noise, which usually consists of white noise. It is impossible to fully eliminate background noise due to its extremely wide spectrum. To remove all of it, you would have to delete the entire signal.

Instead, denoisers target the most prominent components, such as whistling, buzzing, camera motor noise and much more. These parts of the signal are attenuated while preserving as much of the rest of the signal as possible.

Denoisers usually work by learning. You need to provide a representative sample of the noise that needs to be eliminated as an input, from which the denoiser generates a sound profile. This profile is then used to target and attenuate noise in the rest of the audio signal. There is often also a second smoothing phase to reduce any artifacts that may have been introduced by the first phase.

9.3.4. *Declippers*

Declippers fix clipping issues caused by saturation in the recording or overly ambitious analog-to-digital conversion.

The clipped waveforms are replaced with new synthesized and recalculated audio data in order to remove any flat regions in the waveform of the saturated signal.

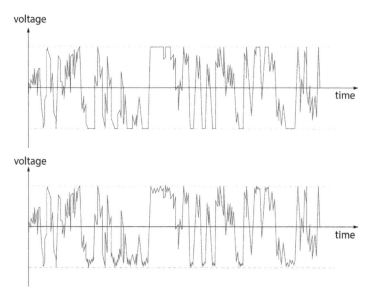

Figure 9.12. *Top, a clipped audio signal. Bottom, the same signal after processing with a declipper*

9.3.5. *Debuzzers*

Debuzzers are more specific and targeted versions of denoisers designed to eliminate hissing and buzzing. They usually target the fundamental frequency of the unwanted noise, and then attempt to predict which harmonics might also need to be deleted.

REMARK.– This tool is sometimes called a *dehummer*.

9.3.6. *Examples of restoration tools*

Table 9.2 lists a few restoration tools. Today, most restoration tools on the market are software based, often sold as suites with multiple features.

Sound restoration racks are uncommon, but some manufacturers do offer a few highly effective models.

Figure 9.13. *Sound restoration racks (from top to bottom): three racks by Cedar, the "DNA-1" model by Weiss and a very popular model, the "SNR2000" denoiser by Behringer*

Type	Manufacturer or publisher	Model	Remarks
Rack	Behringer	SNR2000	Denoiser – Stereo
	Behringer	SNR202	Denoiser – 2 channels
	Behringer	SNR208	Denoiser – 8 channels
	Cedar	CRX Decrackler	Decrackler
	Cedar	DCX Declicker	Declicker
	Cedar	BRX Debuzzer	Debuzzer
	ELP	Declicker	Vinyl declicker (HiFi) – Stereo
	Weiss	DNA1	Denoiser/declicker/decrackler – Ambient sound processor – M/S encoder/decoder
Software	Accusonus	Era-D	Denoiser and reverb suppressor Windows: VST2; VST3; AAX OSX: VST2; VST3; AU; AAX
	Acon Dgital Media	Restoration Suite	Denoiser/declicker/declipper/dehummer Windows: VST; VST3; AAX OSX: VST; VST3; AU; AAX
	Cedar	Cedar Studio	Declicker/debuzzer/decrackler and more Windows: VST; RTAS OSX: VST; RTAS; AU
	Izotope	RX6 Advanced	Windows: VST2; VST3; AAX; RTAS OSX: VST; VST3; AU; AAX; RTAS; DPM Standalone version available

Izotope	RX6 Standard	Windows: VST2; VST3; AAX; RTAS OSX: VST2; VST3; AU; AAX; RTAS; DPM Standalone version available
Magix	SOS Vinyl and K7 Audio	Software solution for digitizing vinyls and K7s with integrated restoration features Microsoft Windows Version
Sonic Solutions	NoNoise3	Denoiser/dehummer/debuzzer and more Windows: VST OSX: VST; AU
Sonic Studio	Legendary Audio I.C.E. (In Case of Emergency)	Audio restoration for very low quality recordings Windows: VST; AAX OSX: VST; AAX; AU
Sonnox	Restore	Denoiser/debuzzer/declicker Windows: VST2; AAX; RTAS OSX: VST2; AAX; AU; RTAS
Sonnox	Oxford Declicker Native	Declicker Windows: VST2; AAX; RTAS OSX: VST2; AAX; AU; RTAS
Sonnox	Oxford Debuzzer Native	Debuzzer Windows: VST2; AAX; RTAS OSX: VST2; AAX; AU; RTAS
Sonnox	Oxford Denoiser Native	Denoiser Windows: VST2; AAX; RTAS OSX: VST2; AAX; AU; RTAS
TC Electronic	Restoration Suite	Denoiser/declicker/descratcher/decrackler/dethumper[3] Windows: VST; AAX; RTAS OSX: VST; AU; RTAS
Wave Arts	Master Restoration Suite	Denoiser/declicker/dehummer and more Windows: VST; AAX; RTAS OSX: VST; AU; RTAS
Waves	X-Click	Part of the "Restoration" suite
Waves	Z-Noise	Part of the "Restoration" suite
Waves	X-Hum	Part of the "Restoration" suite
Waves	X-Noise	Part of the "Restoration" suite
Waves	X-Crackle	Part of the "Restoration" suite
Waves	Restoration	Declicker/denoiser/decrackler/dehummer Windows: VST2; VST3; AAX OSX: VST; VST3; AU; AAXFF

Table 9.2. *Examples of audio restoration tools*

---

[3] Elimination of low-frequency pulses.

9.3.7. *Final remarks on sound restoration*

The tools presented in this section are not the only tools that you might need in order to properly restore sound. There are a host of other problems that you might encounter, such as poor dynamics, loss of treble frequencies, low-quality stereo and so on.

To solve these issues, you will need to use tools such as compressors (see section 7.2), equalizers (see section 4.2), enhancers (see section 7.3) and many others. The objective is to fully reconstruct the original sound message in order to achieve a high-quality and natural sound rendering that is true to the original source.

All of these tools are presented in the previous chapters. Do not hesitate to flip back and forth as much as you need.

9.4. Loopers

Loopers can be used to repeat a certain musical passage *ad infinitem*, but also have other applications.

This effect has been around for a long time. The first implementations used tape recorders, later followed by samplers, and finally by microcomputers, which are still used today. However, loopers in the form of pedals are much more convenient and user-friendly for practical situations.

Although loopers are most commonly employed by guitarists, they can be used with any instrument, and even microphones are (sometimes) compatible with this type of pedal.

Figure 9.14. *Three looper pedals (from left to right): The "JamMan Looper" model by Digitech, the "Ditto Looper" model by TC Electronic and the "22500" model by Electro Harmonix*

The duration of each loop depends on the size of the internal memory of the pedal, as well as the quality of the recording. They usually allow large or sometimes unlimited numbers of loops (*overdubs* or re-recordings) to be layered on top of each other.

The latest generations of loopers offer a variety of features, including effects (reverse, delay, etc.), integrated rhythms, metronomes, presets, etc.

9.4.1. *Looper connections*

There are several possible ways to connect a looper in the audio flow. Unlike some other effects, loopers apply a very specific processing operation, and so their position is critical.

The simplest solution is to place the looper between the instrument and the amp, like most other effects.

But if you need to chain multiple effects together, you will need to decide whether to place your looper before, after or between the other pedals.

9.4.1.1. *Before other effect(s)*

If you place other effects after the looper, you need to remember that any effects that are active when the looper is used will be present in the loop. This might not necessarily be desirable. For example, if you are using a fuzz pedal, your loop will quickly become chaotic.

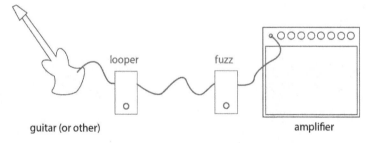

Figure 9.15. *A looper pedal placed before a fuzz pedal*

9.4.1.2. *After other effect(s)*

If we switch the order of the two pedals, only the sound of the guitar is processed when the fuzz pedal is turned on. The loop remains neutral and clear.

Of course, this assumes that the loop was previously recorded with the fuzz pedal turned off.

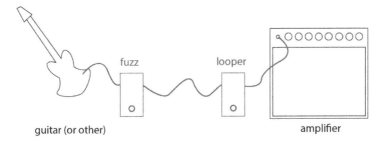

Figure 9.16. *A looper pedal placed after a fuzz pedal*

This example clearly shows the importance of the order of the looper and fuzz pedals. However, in other cases, for example with slight reverb, it might be preferable to keep the effects pedal after the looper pedal.

9.4.1.3. Between effects

As you might have guessed, placing a looper between two or more other effects requires some thought. Any effects placed after the looper will always be applied to the prerecorded loop, whereas the effects placed before it can be used independently from it.

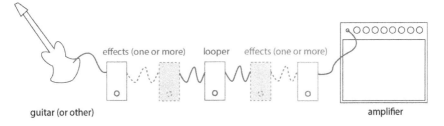

Figure 9.17. *A looper pedal placed between several other effects*

9.4.1.4. In an effects loop

Some amplifiers include a so-called effects loop, similar to those found on mixing desks. The effects loop allows something else to be placed between the pre-amp and the power amp.

The loop usually has two connections called "SEND" and "RETURN".

274 Musical Sound Effects

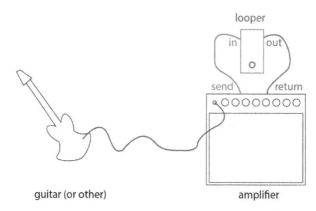

Figure 9.18. *Connecting an instrument and a looper pedal to an amp with an effects loop*

The SEND on the amp needs to be connected to the input (IN) of the looper pedal with one cable, and the output (OUT) of the looper pedal needs to be connected to RETURN on the amp using a second cable, as shown in Figure 9.18.

This setup has several advantages. Any modifications applied to the pre-amp, such as overdrive, channel settings and internal effects (tremolo, reverb, etc.) will be ignored by the looper, which simply reproduces the originally recorded loop.

9.4.2. *Examples of looper pedals*

Manufacturer	Model	Remarks
Boss	RC-505 Loop Station	Five stereo tracks – 99 memory slots – MIDI controls – effects – live or studio – advanced features
Boss	RC-30 Dual Track Looper	Two stereo tracks – 99 memory slots – 3 h of stereo recording
Boss	RC-1	Stereo – 12 min of recording – Very simple
Digitech	JanMan Stereo	Stereo – 35 minutes of recording – 16 h of recording to SD card – 99 memory slots – advanced features
Electro Harmonix	22500 Dual Stereo Looper	Stereo – 12 h of recording (SHDC card) – 100 loops per card – 16 integrated percussions loops – effects – advanced features
Electro Harmonix	2880 Super Multi-Track Looper	Two stereo tracks or 4 mono tracks – Compact flash card – 31 min of recording with a 1-GB card – MIDI – simple to use

Electro Harmonix	Nano Looper 360	Six minutes of recording – 11 memory slots – unlimited overdubs*
Line 6	JM-4 Looper	Stereo – integrated tuner – presets – effects – equalizer – 100 tracks – Mic input – memory card – advanced features
Mooer	Micro Looper	30 min of recording – unlimited overdubs* – very simple
TC Electronic	Ditto Looper	Very simple – 5 min of 24-bit recording – one track – unlimited overdubs*
TC Electronic	Ditto X2 Looper	Five minutes of 24-bit recording – unlimited overdubs* – loop imports and exports – effects
TC-Helicon	Ditto Mic Looper	Loop pedal for microphone – 5 minutes of 24-bit recording – XLR input/output – phantom power

*Re-recordings.

Table 9.3. *Examples of looper pedals*

9.5. Time stretching

This effect modifies the tempo and thus the duration of a recording by making it longer or shorter in order to synchronize it with another audio signal without changing the pitch.

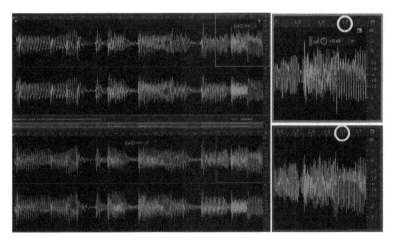

Figure 9.19. *An audio signal with a duration of 2.1 s (top) after applying 1.5× time stretching, resulting in a duration of 3.15 s (bottom). The red and blue boxes show the equivalence between time scales: 2.0 s -> 3.0 s (in yellow) (the editor is Adobe Audition CC). For a color version of this figure, see www.iste.co.uk/reveillac/soundeffects.zip*

Most time stretching effects use one of three main techniques called "phase vocoder", "spectral sine modeling" and SOLA (Synchronous OverLap and Add), which we will not discuss in detail here (see the web links at the end of the book).

Many audio editors and DAWs offer at least one of these three techniques.

REMARK.– Features like *"pitch scaling"* are often associated with time stretching, allowing an audio signal to be transposed without changing its duration (constant speed). Some harmonizers (see section 6.2.3) also include these features.

9.6. Resampling

Resampling is strictly speaking more of a processing operation than an effect. It modifies the bit depth or sampling frequency of a digital audio signal to make it compatible with a different medium, for example converting from 24-bit sound in a DAW to a 16-bit signal for CD, or converting from a 44.1-kHz CD to a 48-kHz K7 DAT.

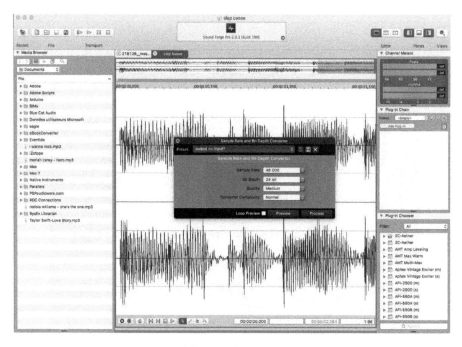

Figure 9.20. *Resampling a 48-kHz 24-bit audio file with the software editor Sound Forge Pro. For a color version of this figure, see www.iste.co.uk/reveillac/soundeffects.zip*

9.7. Spatialization effects

To expand on sections 2.6 and 2.7 of this book, we will now list a few examples of spatializers with relatively specialized features. However, in my personal opinion, fully mastering effects such as compression, reverb, delay, equalization and filtering is the most important step toward properly understanding spatialization, especially for the production of conventional mixes.

Figure 9.21. *IRCAM SPAT 3, a very powerful spatializer. For a color version of this figure, see www.iste.co.uk/reveillac/soundeffects.zip*

These tools can be useful to simulate the sound environments of specific locations, or even artificial sound environments, for example in the context of video games.

Publisher	Name	Remarks
Auburn Sound	Advanced Binaural Panner	Freeware Windows: VST OSX: VST; AU
Taucher interact	MNTN The Sound of the Mountain	Windows: VST; VST3 OSX: VST; VST3; AU
Longcat	Binaural Spatializer	Windows: VST
Flux	Ircam SPAT V3	Windows: VST; AAX OSX: VST; AAX; AU
Tokyo Dawn Records	Proximity	Freeware Windows: VST OSX: VST; AU

Table 9.4. *Examples of dedicated spatializer plugins*

9.8. Conclusion

The effects and processes assembled in this final chapter are those that seem most important to me personally, but there are many more that we do not have the time or space to mention. To find out more, interested readers can visit the websites of each publisher and product brand. Downloading user manuals and instructions, often available free of cost, is another great way to find more information.

Conclusion

After nine chapters and four appendices, we have finally reached the end of this book. I hope that reading through these chapters and appendices has allowed you to discover, explore and understand the fascinating universe of digital sound effects.

No doubt there is much left to say. Many effects other than those mentioned here would arguably have deserved a place within these pages, but my own knowledge is far from perfect. I can only talk about effects with which I am personally familiar – the ones that I have encountered and used myself over the course of my professional career.

I work with some of these effects every day in my studio, but even after all these years, I still discover new settings, approaches, combinations and configurations every single day. These discoveries allow me to build ever more striking and attractive sounds.

Working at my mixing desk and workstation is never a chore. Nothing gives my eyes and especially my ears more pleasure than listening, relistening, tweaking and perfecting the musical passages that have been entrusted to me. In my work, I always try to satisfy the needs of the people for whom the sounds are ultimately intended – whether for live performances, radio, sound reinforcement, artistic creativity, digital art or even just family.

I also love to trade stories with performers, composers, sound professionals and anyone else who shares my passion and eagerness for working with sound.

Since the late 1990s, it has become increasingly rare for completely new effects to be invented. But the digital age was nonetheless an earth-shattering revolution for the world of audio. The sophistication of DAWs and audio editors has exploded.

Audio software programs have forged increasingly close connections with their users, taking advantage of the rapid expansion of computing power from year to year.

Internet and networking technology has taken root, improving discussions, simplifying cabling problems and standardizing the various components of audio systems as much as realistically could have been possible.

Today, there is software for almost everything. Even the oldest effects, created in the 1960s, have been republished, and in some cases improved and reinvented.

My personal opinion is that you do not need dozens of different plugins. Instead, choose the characteristics of the plugin carefully to reflect what you are trying to achieve. Do not hesitate to download trial versions to try out any products that you find on the market so that you can form your own opinion.

Similarly, I believe that digital solutions are not always the answer, though they are unquestionably capable of amazing things. Personally – and many others share my views on this – nothing can beat a good old studio rack or hardware versions of classic effects, even if they are dinosaurs in terms of bulk and usability. If you are not already convinced, see if you can find an opportunity to listen to a Fairchild 670 compressor, a Binson echo chamber, a Pultec EQP-1A equalizer, an MXR 100 phaser, Leslie speakers or any other equipment of similar pedigree. The experience is well worth your while. By placing any of these hardware effects at the right position in the audio flow, you can create wonders – these effects give you the physical sensation of shaping the audio signal with your own hands, your own body. The virtual perception of digital effects simply cannot compare with this.

Of course, the flip side is that hardware equipment is often prohibitively expensive. This is the price of perfection in sound rendering.

However, it is often said that no man is a prophet in his own country, and I do not doubt that this nugget of wisdom applies to me. I consider myself lucky to have started working with sound decades ago, but I will undoubtedly have settled into routines and habits over the years, both good and bad.

What might the future look like? We can expect our equipment to improve even further. Network bandwidth and transfer speeds will continue to increase, more advanced software will be developed. Real-time sound processing will also continue to mature. These things are hardly difficult to foresee. But who knows what surprises might lurk around the corner?

One domain that might experience significant change in the not-too-distant future is the reproduction and playback of new virtualized musical formats. Even though the multichannel systems available today are extremely advanced, our ears can still distinguish between real and virtual sound environments. This is one aspect of sound rendering that potentially still offers some room for improvement.

Finally, it seems that a slight vintage breeze has begun to blow throughout the small world of sound recording. Vinyls are making an unexpected comeback. Old analog effects seem to be regaining some momentum. Is this just a passing trend, or does it perhaps represent a more lasting shift in preferences? Time will tell.

Appendices

Appendix 1

Distortion

A1.1. Introduction

Distortion is often denounced as the public enemy number one of sound reproduction. But some musicians create distortion deliberately to create certain colors for the sound of their instruments. What does this mean, and what is the true nature of distortion?

A1.2. Measuring distortion

To quantify the quality of sound production equipment, we often measure the distortion that it generates. When doing so, it is important to take account of the most relevant type of distortion – there are several.

Distortion is more than just a single isolated concept. There are various types of distortion described in terms of six parameters that we can measure independently:

– the level;

– the frequency response;

– the signal-to-noise ratio (SNR);

– the harmonic distortion;

– the cross-talk;

– the phase.

A1.2.1. *The level*

This parameter describes the level of the output signal relative to the input signal. The difference between the two is the distortion. This might seem like an unusual definition at first, but distortion is conceptually an alteration of the original signal, and so variation in the volume may be viewed as a form of distortion.

A sound source is said to be distortion free if the reproduced sound level is equivalent to the original volume of the sound when it was emitted.

But in practice, we are perfectly used to recreating sounds much louder than their original input volume, and so this does not seem unnatural to us.

When a signal is amplified, an enormous amount of energy is lost by the equipment (cables, pre-amp, amp, speakers, etc.). Most of this is due to the Joule effect (the resistance of the components creates losses in the form of heat, among other factors).

When an amplification system is used to increase the volume of a sound, the system defines a nominal level for 0 dB. If this value is exceeded, the capacity of the equipment is overwhelmed, and the sound is altered, often in the form of saturation.

The exact nature of this saturation varies according to the technology on which the equipment is based: tube amps, transistors, integrated circuits or hybrid amps, with classes A, AB, B, D, H, etc.[1]

To measure the level, we apply an audio signal input with suitable characteristics, then increase the volume up to the nominal value before taking a measurement.

The result is a sound pressure reading expressed in either dB SPL (sound pressure level) or pascal (Pa).

94 dB SPL is equivalent to a sound pressure of 1 Pa.

$$L_p = 20 \times log_{10}\left(\frac{P_{eff}}{P_{ref}}\right) \text{ in dB}$$

---

[1] See Appendix 2.

where:

- L_p: sound pressure in Pa;
- P_{eff}: effective value;
- P_{ref}: reference sound pressure (20 µPa).

If the sound cannot be physically measured by recording using a microphone, we can instead measure the electrical voltage at the terminals of the output (headphones, earphones, amp, etc.) and compare the results with the voltage of the input signal.

A1.2.2. The frequency response

Depending on the type of hardware used by the sound reproduction system, the sound level can vary as a function of the input signal frequency. This is often a key factor in the characterization of the sound, determining whether the sound seems warm, crystalline, messy, clear, precise, sharp, flat, neutral, etc.

It is extremely uncommon for any given piece of equipment to be perfectly linear over the whole audible spectrum (20–20,000 Hz).

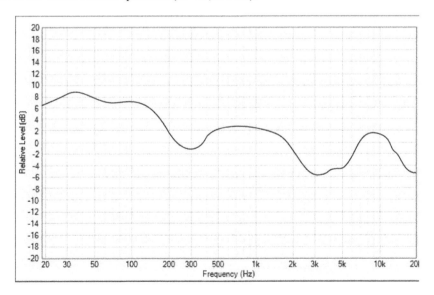

Figure A1.1. *Example of a frequency response curve*

Measuring the frequency response can be tricky, since it depends on many different parameters.

To measure the frequency response of a speaker, the usual approach is to record with a microphone. But this microphone is itself never perfectly linear, and the location in which the measurement is being executed can also affect the results (reflections, damping, etc.). Another factor is the distance between the microphone and the source – there are many others.

By contrast, the frequency response of amplifiers is measured by examining their output signal when a load is applied (typically 4, 8 and 16 Ω). The load is chosen to imitate reality as best as possible.

One often-quoted property is the global bandwidth of the system, for example 30–18,500 Hz. The usual definition of the global bandwidth is the frequency band on which all sound signals have a level of at least 3 dB[2].

A1.2.3. *The signal-to-noise ratio*

The SNR is an extremely important factor when evaluating the quality of a sound system.

Many auxiliary signals can potentially interfere with the sound reproduction system. The choice and nature of the electronic components in the system, the shielding of the power supply, the quality of the cables and many other factors all affect the results to some degree or another.

Each of these factors can interfere with and distort the outgoing audio signal by introducing background noise.

This background noise often takes the form of a permanent hissing that attaches itself to the sound content. Unwanted noise can obscure some of the information contained by the original input.

Electronics defines five types of noise that can make up undesirable hissing: thermal noise (resistance noise or Johnson noise), shot noise (or Poisson noise or Schottky noise), flicker noise (or low-frequency noise or excess noise), avalanche noise and burst noise (random telegraph signal (RTS) noise or popcorn noise or impulse noise):

– thermal noise is caused by thermal agitation of electrons;

---

2 Other definitions of values are sometimes used.

– shot noise is caused by the fact that the electric current is not continuous, but instead consists of individual electrons carrying elementary charges;

– flicker noise is linked to electronic components and is always present whenever they are used. It is often caused by impurities in the constituent materials;

– avalanche noise arises from semiconductors and is caused by the electric field, which accelerates certain electrons;

– burst noise is caused by the creation of voltage peaks of a few microvolts, especially in operational amps[3].

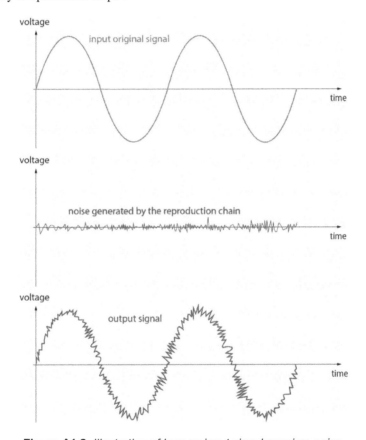

Figure A1.2. *Illustration of how an input signal acquires noise*

---

3 Electronic amplifiers that boost the electrical voltage of its input(s).

To understand the concept of SNR, we need to introduce the notion of dynamic range, which is defined as the ratio between the strongest and weakest sounds that can be emitted by the sound reproduction system.

If the SNR is larger than the dynamic range, everything is fine. If not, things do not look quite so good – low-intensity information is degraded, or even destroyed and lost, drowned out by noise.

The SNR therefore needs to be as high as possible to provide a good listening experience.

Measuring the SNR of a system is straightforward. A perfectly known input signal (S) described in terms of the frequency and intensity is applied to the system, and then compared to the output in order to isolate any additions or changes (N). The difference between the input and the output gives the SNR (S/N_{dB}).

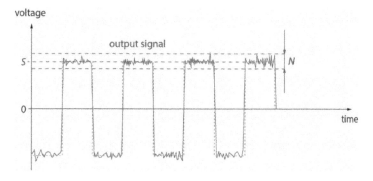

Figure A1.3. *Representation of a signal S partially obscured by noise N. For a color version of this figure, see www.iste.co.uk/reveillac/soundeffects.zip*

A1.2.4. *Harmonic distortion*

Harmonic distortion is also known as the Total Harmonic Distortion (THD). To explain this concept, we first require a definition of the concept of harmonics.

A harmonic is a component of a periodic sound whose frequency is an integer multiple of the fundamental frequency of the signal. Harmonics can be classified into even harmonics (multiples of 2) and odd harmonics.

Harmonic distortion is said to arise within a sound reproduction system whenever a non-negligible quantity of harmonics that were not present in the original input signal are generated by the system. These harmonics represent a distortion of the original sound.

Harmonic distortion is very low in most modern equipment, often completely covered by the background noise measured in terms of the SNR. Therefore, with modern equipment, a parameter called the THD + noise (THD-N) is defined in order to determine whether the harmonic distortion is stronger than the background noise, and, if so, by how much.

The THD is calculated using the following equation:

$$THD = \sqrt{(1^{st}\ harmon.)^2 + (2^{nd}\ harmon.)^2 + \cdots + (N^{th}\ harmon.)^2}$$

For example, with the following input signal, where F denotes the fundamental frequency:

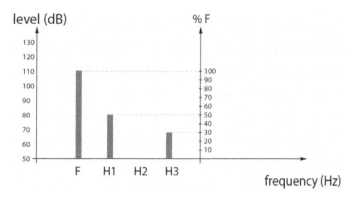

Figure A1.4. *Input signal*

and the following output signal:

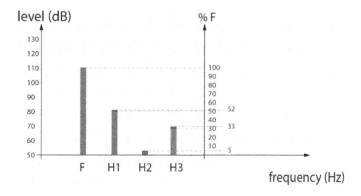

Figure A1.5. *Output signal with distortion (in red)*

the THD is:

$$THD = \sqrt{(52\% - 50\%)^2 + (5\%)^2 + (33\% - 30\%)^2} \approx 0.501 \text{ (i.e. } 5.01\%)$$

The THD can also be expressed in dB using the following transformation formula:

$$THD_{dB} = 20\log\left(\frac{THD\%}{100}\right)$$

In our example, this gives:

$$THD_{dB} = 20\log\left(\frac{5.01}{100}\right) \approx -26.003 \, dB$$

The value of –26 dB means that the intensity of the THD is 26 dB lower than that of the fundamental frequency F.

A1.2.5. *Cross-talk*

Cross-talk measures the tendency of parts of the signal from the x-channel to bleed over into the y-channel when both channels are processed by the same equipment (often the case for stereo sound). This usually arises from phenomena related to insufficient shielding, components placed too closely together on the same circuit, capacitance effects and so on.

High cross-talk means that the signals on multiple channels merge together, resulting in more of a mono signal, losing the stereo or multichannel depth of the original signal.

To measure cross-talk, we apply an input signal to one of the channels, for example the left channel of a stereo system, and then observe whether there is any output on the right channel. In multichannel systems, this procedure is repeated for each channel.

A1.2.6. *Phase*

There are several phase interaction phenomena that can occur whenever you have more than one signal (see section 1.6.2). These phenomena have the unfortunate tendency of destroying parts of each signal.

For example, if you send the same 1-kHz sine wave to both channels of a stereo playback system, you should ideally ultimately obtain two signals that are fully aligned (in phase), which combine to give the same signal with a higher amplitude (the sine waves add together).

But if there is a delay on one of the channels, the crests and troughs of the sine waves will be shifted relative to each other. The resulting sound will still have the same frequency, but will not be so loud. We say that partial phase cancellation has occurred.

If one of the channels is delayed so much that the troughs of one wave align with the crests of the other (and vice versa), the two waveforms cancel fully, resulting in no sound at all. The two signals are in antiphase, and we say that total phase cancellation has occurred. Alternatively, we say that the signals are "180° out of phase".

This cancellation process also occurs if the polarity of one of the channels is inverted. Because of this, people often conflate the two separate phenomena of polarity inversion and phase inversion. They are not the same.

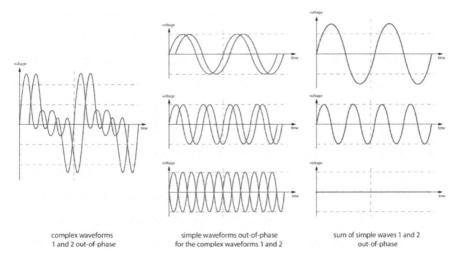

Figure A1.6. *Transformation of a signal composed of two complex out-of-phase waveforms (red and green). For a color version of this figure, see www.iste.co.uk/reveillac/soundeffects.zip*

In practice, the problems created by phase issues are much more complex. Imagine for example that you are recording a guitar passage on two channels. Any delay on one of the channels will in fact change the pitch of the mixed signal instead of its level. Indeed, for a given delay time, the phase relations between the various sine waves[4] on the first and second channels depend on the frequency of each of the wave components in the signal generated by the guitar. If the delay is 0.5 ms, any 1-kHz waves will be in antiphase (1-kHz sine waves have a period of 1 ms). But if the sound of the guitar also includes any components with frequencies of 2 kHz, these components will have a period of 0.5 ms, and so will remain perfectly in phase.

As a result, the intensity of each signal component present in the signal increases as its frequency ranges from 1 to 2 kHz. This is an example of partial phase cancellation, and the sound intensity is highest at the frequency that experiences phase alignment, namely 2 kHz.

This phenomenon is repeated at every higher frequency. The signal will also partially cancel at some higher frequencies, leaving only signal components that are in phase.

With this explanation, you can easily imagine the kinds of problems that can arise from distortion caused by phase shift in sound reproduction systems.

A1.3. Conclusion

There is still much more that could be said about the various types of distortion and the phenomena associated with them. You can find out more by visiting the bibliography and web links at the end of the book.

---

4 Always bear in mind that any audio passage, no matter what kind, is in fact a more or less complex combination of a potentially infinite set of sine waves (as shown by the French mathematician Joseph Fourier, 1768–1830).

Appendix 2

Amplifier Classes

A2.1. Introduction

There is a classification for electronic amplifiers that allows us to characterize them, with one letter for each category of amplifier.

There are 10 major classes: A, AB, B, C, D, E, F, G, H and S. There are also other smaller classes, e.g. the T class, which are usually variants of the major classes.

The first generation of amplifiers was divided into four classes, A, B, AB and C, according to the polarization of their electronic tubes. This same principle was used for the next generation of transistor amps. As new technologies were developed, additional classes were added to the list.

Some classes of amplifier are not relevant to studio work or sound reproduction and playback. These applications are only interested in amplifying signals located in the frequency band from 20 and 20,000 Hz.

Therefore, only classes A, AB, B, D and H fall within the scope of this book.

A2.2. Class A

Most specialists view A as the most faithful and musical class of amplifiers.

High-end hi-fi amplification systems are often based on class A technology, providing great dynamics and excellent musical rendering.

Class A amplification components are 100% active even when there is no signal (at rest). This requires electronic systems capable of dissipating large quantities of heat (radiators, fans and other cooling mechanisms).

Although they run very hot, class A amplifiers are highly stable, with very low distortion rates. The price of these qualities is mediocre energy efficiency; these amplifiers require large power supplies and consume a lot of energy.

Without going into any further detail, there are various subcategories within class A, including "pure class A", A1, A2 and so on.

A2.3. Class B

The amplification components in this class operate at 50% when the amp is at rest. This is more efficient than class A amplifiers, but is slightly less faithful, resulting in more distortion. Today, very few manufacturers offer amplifiers from this class, even though they are less expensive to produce, largely because their power supply does not need to be as robust as class A amplifiers.

A2.4. Class AB

This class is a combination of classes A and B. At rest, the system operates at a level somewhere between 50% (class B) and 100% (class A). When active, the amp gradually transitions from class B to class A as the current increases. Many amplifiers, especially those used by sound reinforcement systems, are based on this technology, which is more efficient than class A, although slightly less responsive.

The power supplies of class AB amplifiers are more reasonably sized than those of class A amplifiers, offering a good compromise.

A2.5. Class D

You might be forgiven for guessing (and indeed some sources mistakenly claim) that the letter D standards for "digital". But class D amps are in fact also analog, not digital.

Class D amplifiers are based on sophisticated technology with maximum efficiency and very low power dissipation. This technology is most suitable for small amplifiers and is perfect for mobile or portable devices.

In the past, these amps were criticized for having poor musicality, but a lot of progress has been made over the last few generations. Many home theater and hi-fi systems now use class D amplifiers.

Class D is also used by many subwoofers and wireless speakers.

A2.6. Class H

To improve the efficiency of a class B or AB amplifier, one idea is to vary the voltage as a function of the audio signal. Without going into too much detail, this is how class H systems work.

The main advantage is significant improvements in energy efficiency, which means that a much smaller power supply is required. Sadly, this requires a sacrifice in audio quality. This technology creates distortion in some usage conditions, especially in intermediate volume regions, where the power supply voltage can be expected to vary a lot.

Class H amplifiers achieve excellent energy efficiency and can also be made very compact, since they only require a small power supply.

A2.7. Conclusion

This appendix has hopefully helped to clarify the meaning of the various classes of amplifier. I deliberately did not attempt to go into extensive detail or approach the topic of digital amplifiers (binary switching amplifiers). If you are interested in a more in-depth discussion of the electronics, design, measurements and performance of amplifiers, you can visit the bibliography and the web links at the end of the book.

Appendix 3

Introduction to Max/MSP

A3.1. Introduction

Throughout this book, we have discussed a wide range of ideas about audio signal processing, mostly from a theoretical perspective. Putting these ideas into practice and listening to how they work in practice is an instructive exercise.

Max/MSP by Cycling'74 is an ideal example of a software tool that allows you to do this, although there are plenty of others, both freeware and commercial, that are also perfectly suitable.

Max[1] is my first choice because I have now been using it for more than 30 years. At the time, it was truly revolutionary. We were working on the first generations of Apple computers (Mac Plus, Mac II, Mac SE30, etc.). Since then, Max has changed a lot, but has never lost sight of its guiding principles, which are still perfectly recognizable in the latest versions (Max 7).

REMARK.– "Max for Live" is a version of Max designed to be compatible with Ableton Live 9.

The concepts presented here are easy to adapt to other software tools (Pure Data, SuperCollider, jMax, etc.). You will find everything you need at the end of the book in the bibliography and web links (textbooks, download links, publishers' websites, forums, tutorials, etc.).

---

[1] Music software for synthesizing sound, controlling instruments, recording, performing audio analysis and more. Max was developed in France by IRCAM (Institute for Research and Coordination in Acoustics/Music) in the 1980s.

This appendix simply introduces a few basic examples with Max/MSP that will allow you to start experimenting for yourself. If you are interested, you can then keep exploring further on your own.

If you asked me the question: "What can you do with Max/MSP?", my answer would be: anything, or almost anything that you could possibly imagine for audio or possibly even video. Max/MSP is a vast toolbox, almost like a more user-friendly version of a programming language. It has a completely open structure and supports connections with many other software programs, robots, electronic instruments, smartphones, etc.

A3.2. How Max works

Max works almost exclusively with graphical interfaces. Blocks (boxes or elements) are placed into a window and connected together. Each block has its own role, with inputs and/or outputs, and an array of configurable parameters that can be defined and edited.

Each set of connected blocks is placed in a window called a "*patcher*".

A3.3. Getting started: sine wave oscillator

For our first project, we will construct a small sine wave generator that can be manually varied from 20 Hz to 20 kHz.

Open Max, go to the FILE menu, and select NEW PATCHER. The window that opens is now your main window.

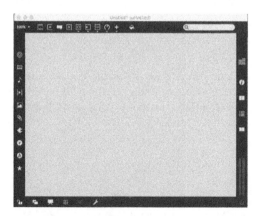

Figure A3.1. *The main window of MAX (version 7)*

Appendix 3   301

The top toolbar allows you to select blocks. They are listed separated or grouped into families. From left to right: objects, message, comment, toggle, buttons, numbers, sliders, Max for Live objects and add object.

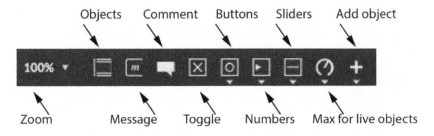

Figure A3.2. *The horizontal toolbar*

REMARK.– The 100% dropdown in the top left is for adjusting the scale (zoom ±).

When you click on OBJECT, a block is created in the center of the window. You can move this block by dragging it.

An object is a box whose function has not yet been configured. When a name is entered in the box, the object gains a function.

The blinking text cursor indicates that the software is waiting for you to type the name of a primitive. For our example, we will choose: "cycle". Immediately after starting to type, autocomplete pops up, allowing you to select "cycle~" (sinusoidal oscillator), as shown in Figure A3.3.

This creates a sine wave oscillator.

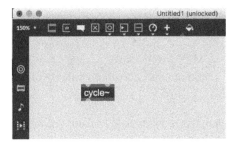

Figure A3.3. *An OBJECT block with the "cycle~" function, i.e. a sinusoidal oscillator*

Now, click on the NUMBERS family and select the first block: "number".

A number stores an integer or floating-point number value for a parameter of an object.

These steps create a tool for configuring the frequency of the oscillator. We will configure the pitch so that it can be varied in steps of 1 Hz.

Move the number to just above the oscillator by dragging it, then connect the output of the number (bottom left) to the input of the oscillator (top left). To do this, simply click and drag between the two points to draw a connection, as shown in Figure A3.4.

When you click on a connection, you will see two small colored dots. These dots define the direction of the connection – green for start and red for end.

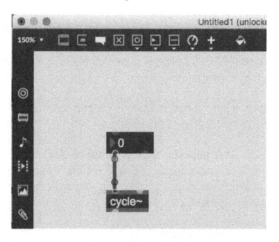

Figure A3.4. *The two blocks are now connected. For a color version of this figure, see www.iste.co.uk/reveillac/soundeffects.zip*

Now, single-click on MESSAGE twice to create two new blocks. We will use these blocks to define minimum and maximum values for the frequency parameter of the oscillator.

The purpose of each message block is to transfer one value or parameter.

Type "min 20" in the first message block and "max 20,000" in the second. Move them so that they are above the number block, then connect them together as shown in Figure A3.5.

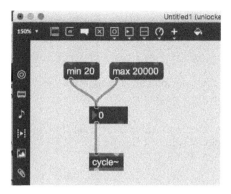

Figure A3.5. *After moving the MESSAGE blocks into position*

Next, click on the ADD OBJECT family, select the AUDIO subfamily and choose the block that is shaped like a speaker (ezdac~). This block creates an amplified audio signal output, which is usually sent to the sound card and speakers of your computer.

Connect this block to the oscillator (see Figure A3.5). We need two connections in order to support a stereo signal on both channels, left and right.

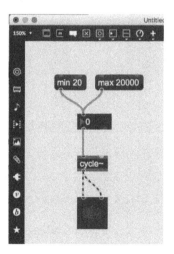

Figure A3.6. *The audio output block is now connected*

Next, open the FILE menu, choose SAVE, type a name for your file, select a file location and finally click SAVE.

We are now ready to test our oscillator. Click on the AUDIO ON/OFF icon at the bottom left of the active window. It should turn blue, and you should see the ezdac~ block turn on.

You may need to lock your patcher by clicking the padlock icon at the bottom left of the active window.

In order to initialize the limit values, click on the "min 20" MESSAGE block, followed by "max 20,000".

Now, click and hold the number block, then move your cursor up or down. You should see the frequency value displayed in the block change, and you should be able to hear how the sound changes accordingly.

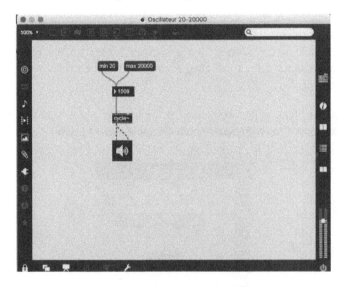

Figure A3.7. *Your patcher, working and locked*

Click on AUDIO ON/OFF to turn your oscillator off.

REMARK.– Locked mode prevents you from making any changes. To activate the audio mode, you can still click on the ezdac~ block.

A3.4. Improving your oscillator

For the next step in our project, we will add a component to our oscillator that allow us to visualize the generated waveform. We will also automate the configuration of the number block. This avoids the need to initialize the min (20) and max (20,000) values manually and ensures that the working range of the oscillator is bounded.

Click on the ADD OBJECT icon, choose the AUDIO subfamily, then select a scope~ block.

Drag the edges of this block slightly so that it becomes a rectangle, then connect a number block and the output of the cycle~ object to its input, as shown in Figure A3.8.

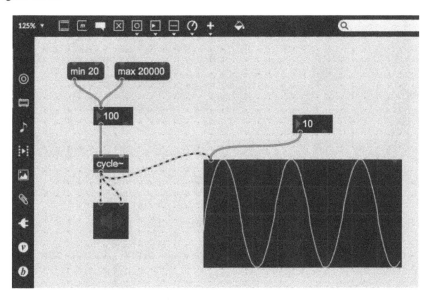

Figure A3.8. *The oscillator with an added graph (oscilloscope) for visualization*

The number block connected to the scope~ block allows us to manage the size of the signal display buffer. Changing the value of the number block will extend or reduce the length of the curve. This parameter is essentially equivalent to the horizontal "Time/Div" setting on physical oscilloscopes.

To automatically configure the working range of the oscillator whenever it is turned on, we simply need to connect a loadbang input to the two message blocks.

Open the FILE menu, select SAVE, then click the SAVE button to save your changes.

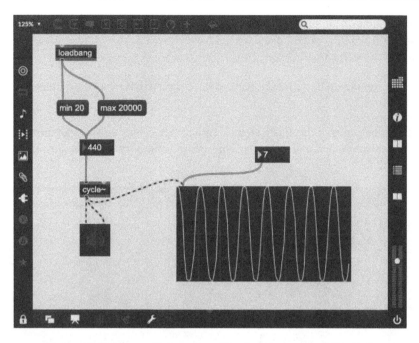

Figure A3.9. *Our oscillator, now with a loadbang object (top left)*

A3.5. Tremolo, vibrato, reverb and chorus

After mastering the simple example presented above, you can start playing around with more interesting patchers.

REMARK.– You can access the help function by right-clicking on any block. This displays one or several working examples of prebuilt, modifiable patchers using this block.

Figures A3.10–A3.13 show four ready-made patches for building, experimenting and listening to tremolo, vibrato, reverb and chorus effects.

REMARK.– In sections A3.5.1–A3.5.4, comments explaining the functionality of each of the implemented blocks and objects are written in italics.

A3.5.1. *Tremolo*

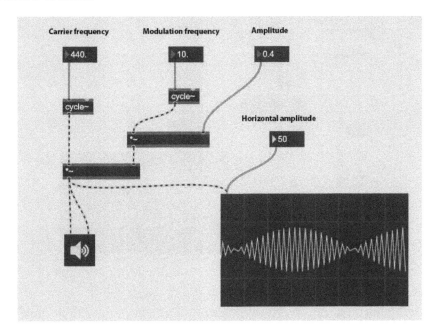

Figure A3.10. *Example of a tremolo*

Tremolo can be constructed using the following blocks:

– three flonum blocks – NUMBERS family – *displays and generates a decimal value*;

– one number block – NUMBERS family – displays and generates an integer value;

– two cycle~ objects – creates a sinusoidal oscillator;

– 2 *~ objects – multiplies two signals together;

– one ezdac~ block – ADD OBJECT family – AUDIO subfamily – *Audio output (on/off)*;

– one scope~ block – ADD OBJECT family – AUDIO subfamily – *visualizes an audio signal*.

REMARK.– You can use "comment" blocks to add text to your patcher.

A3.5.2. *Vibrato*

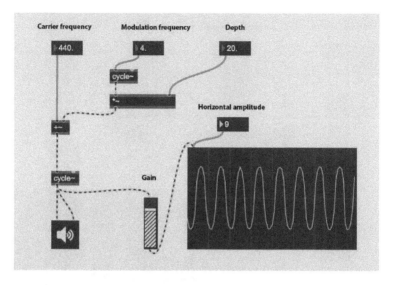

Figure A3.11. *Example of a vibrato*

Vibrato can be constructed using the following blocks:

– three flonum blocks – NUMBERS family – *displays and generates a decimal value*;

– one number block – NUMBERS family – displays and generates an integer value;

– two cycle~ objects – creates a sinusoidal oscillator;

– one *~ object – multiplies two signals together;

– one +~ object – adds two signals together, or adds a compensation value to a signal;

– one gain~ block – SLIDERS family – *controls the gain*;

– one ezdac~ block – ADD OBJECT family – AUDIO subfamily – *audio output (on/off)*;

– one scope~ block – ADD OBJECT family – AUDIO subfamily – *visualizes an audio signal*.

A3.5.3. Reverb

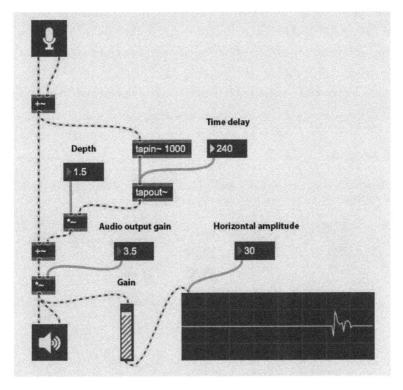

Figure A3.12. *Reverb effect. This effect records the sound of your voice using your computer's microphone (by default)*

Reverb can be constructed using the following blocks:

– two flonum blocks – NUMBERS family – *displays and generates a decimal value*;

– two number blocks – NUMBERS family – displays and generates an integer value;

– two *~ objects – multiplies two signals together;

– two +~ objects – adds two signals together, or adds a compensation value to a signal;

– one gain~ block – SLIDERS family – *controls the gain*;

– one tapin~ object – generates a delay time on the input signal;

– one tapout~ object – generates an output signal with a delay;

– one ezadc~ block – ADD OBJECT family – AUDIO subfamily – *audio input (microphone by default, on/off)*;

– one ezdac~ block – ADD OBJECT family – AUDIO subfamily – *audio output (on/off)*;

– one scope~ block – ADD OBJECT family – AUDIO subfamily – *visualizes an audio signal.*

A3.5.4. Chorus

By removing one number block and adding a few extra elements to the previous patcher (section 3.5.3), we can transform reverb into chorus.

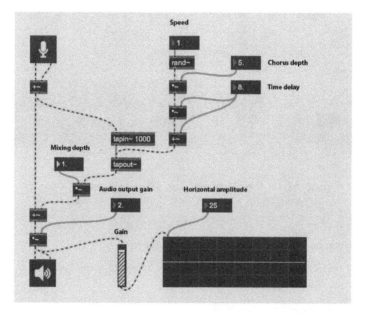

Figure A3.13. *Chorus based on the previous patcher. The chorus records the sound of your voice using your computer's microphone (by default)*

We need to add the following blocks:

– three flonum blocks – NUMBERS family – *displays and generates a decimal value*;

– two *~ objects – multiplies two signals together;

– one +~ object – adds two signals together, or adds a compensation value to a signal;

– one rand~ object – generates a random signal between –1 and 1 of an input frequency.

A3.6. Going further

As mentioned earlier, this appendix is just an extremely basic introduction to Max/MSP. If you would like to build other patchers, you can do so by exploring the documentation available on the Cycling'74 website in the "Max" section, as well as the web links given at the end of the book.

Appendix 4

Multieffects Racks

A4.1. Introduction

This appendix presents Tables A4.1 and A4.2 with a collection of multieffects racks, also known as effects processors, including both historical (vintage) models and more modern examples. Some of these are specifically intended for guitar or bass.

The advantage of these racks is that they combine several dozens of effects into the same device. Some models allow multiple effects to be chained together.

However, the controls of this equipment are not always user-friendly. Managing the settings and sound quality can be tricky. Therefore, you should definitely take the opportunity to try them out if you can, and consider some of the many reviews, tests and demonstration videos that can be found online (YouTube, Daily Motion, Vimeo, etc.) to form your own opinions.

You can also find more information on the manufacturer's website of each model (see Bibliography).

A4.2. Vintage racks

I have done my best to give an approximate release date for each model in this list. Racks are sorted alphabetically by name of manufacturer.

Manufacturer	Model	Release year	Remarks
Alesis	Midiverb	Early 1980s	Twelve bits – 63 presets – Reverb, reverse, echo, MIDI
Alesis	Midiverb II	Early 1980s	Sixteen bits – 99 presets – Reverb, delay, echo, chorus, etc. – MIDI
Alesis	Midiverb 4	1985	Eighteen bits – 128 presets – 128 user programs – Reverb, delay, chorus, flanger, pitch, etc. – MIDI
Alesis	Quadraverb	Early 1990s	Sixteen bits – 100 presets (90 in ROM) – Equalizer, pitch, delay, reverb – MIDI
Boss	SE-50	1990	Twenty-eight presets – 100 user slots – Reverb, chorus, flanger, phase, equalizer, etc. – MIDI
Ensoniq	DP4	Early 1990s	Twenty-four bits – 400 presets including 200 user slots – Reverb, chorus, flanger, delay, distortion, pitch, etc.
Eventide	Eclipse studio	2001	Twenty-four bits – 11 reverbs, delay, chorus, pitch shift, loop – AES/EBU – S/PDIF – MIDI –XLR
Ibanez	UE405	Mid 1980s	Vintage – Analog delay, equalizer, chorus, compressor, limiter
Kawai	RV4	Late 1980s	Fifty presets – 50 user slots – Reverb, delay, chorus, vibrato, tremolo, etc. – MIDI – AES/EBU
Korg	A3	1990	Vintage – For guitar – Compression, distortion, delay, modulation, reverb
Lexicon	LXP15 II	Early 2000s	128 presets – 128 user slots – Reverb, chorus, delay, pitch, etc. – MIDI
Lexicon	MPX-1	1996	Eighteen bits – 200 presets – Reverb, chorus, pitch, equalizer, modulation, delay – MIDI

Lexicon	PCM41	1980	Reverb, delay, vibrato, flanger, vocoder, etc.
Roland	SDX-330	1994	Sixteen bits – 100 presets – 200 user programs – Chorus, flanger, phaser, delay, pitch, rotary, etc. – MIDI
Sony	DPS-V55	1998	Twenty bits – 200 presets – 200 user programs – 45 effects – MIDI
TC Electronic	M-One XL	1999	Twenty-four bits – 200 presets – 100 user slots – Reverb, delay, chorus, flanger, compressor, noise gate, etc.
TC Electronic	M2000	1995	Twenty-four bits – 250 presets – 128 user slots – Reverb, delay, pitch, equalizer, etc. – AES/EBU – S/PDIF – MIDI
TC Electronic	FireworX studio	1998	Twenty-four bits – 400 presets – Vocoder, filter, phaser, special effects, etc. – SLR – MIDI – ADAT S/PDIF
Yamaha	SPX90	1985	Sixteen bits – 30 presets – 60 additional effects – MIDI
Yamaha	SPX900	Late 1990s	Sixteen bits – 50 presets – 49 user slots – Reverb, flanger, chorus, phaser, tremolo, pitch, compressor, distortion, etc. – MIDI
Yamaha	EMP100	1991	Sixteen bits – 100 presets – Reverb, chorus, flanger, delay, pitch – MIDI
Yamaha	SPX990	1993	Twenty bits – 80 presets – Reverb, delay, chorus, flanger, phaser, pitch, etc. – MIDI
Yamaha	FX500B	1990	Sixteen bits – Specially designed for bass – 30 presets – 30 user slots – Compressor, overdrive, equalizer, modulation, reverb, delay – MIDI
Yamaha	EMP700	1994	Sixteen bits – 90 presets – 50 user slots – Equalizer, compressor, limiter, reverb, delay, enhancer, distortion, pitch shift, wah-wah, phaser, modulation, etc.

Yamaha	SPX2000	2003	Twenty-four bits – 80 presets – Reverb, delay, pitch, modulation, gate, etc. – AES/EBU – XLR – MIDI
Zoom	RFX-2000	Early 2000s	Twenty bits – 616 presets – 100 memory slots – Reverb, delay, phaser, pitch, chorus, etc. – MIDI – S/PDIF

Table A4.1. *Examples of vintage multieffects racks*

A4.3. Modern multieffects racks

Table A4.2 gives a list of more recent racks. Most of these are still available for sale at the time of writing. As before, they are sorted alphabetically by name of manufacturer.

Manufacturer	Model	Remarks
Eventide	H8000FW	Twenty-four bits – Multichannel multieffects – 1,600 presets – AES – ADAT – S/PDIF
Behringer	FX2000	Twenty-four bits – 3D effects – Amplifier emulation – Psychoacoustic effects – MIDI
Lexicon	MX300	Twenty-four bits – 16 reverbs, compressor, de-esser, etc. – VST plugin – USB port – MIDI
Lexicon	MX400	Twenty-four bits – 17 reverbs, delay, compressor, etc. – VST plugin – USB port – MIDI
Lexicon	MX200	Twenty-four bits – 16 reverbs, compressor, de-esser, etc. – VST plugin – USB port – MIDI
Lexicon	MX550	Twenty-four bits – 240 presets – Reverb, flanger, chorus, pitch, delay, echo, etc.
Line 6	Helix rack guitar processor	Twenty-four bits – For guitar – Amplifier + effect simulation – S/PDIF – AES/EBU – VDI – USB

Line 6	POD HD Pro X	For guitar – 512 presets – 128 user programs – 22 amp simulations, 19 delays, 23 modulation effects, 17 distortion effects, 12 compressors, 26 filters, 12 reverbs – USB – MIDI – S/PDIF – AES/EBU
Peavey	Dual Deltafex	Twenty-four bits – 200 presets including 100 user presets – Delay, reverb, modulation, distortion, rotary, chorus, etc.
Rocktron	Xpression	Twenty-four bits – For guitar and bass – 128 presets – Reverb, pitch shifting, compressor, delay, chorus, flanger, rotary, phaser, etc.
TC Electronic	M3000	Twenty-four bits – Reverb + effects – Numerous presets – S/PDIF – AES/EBU – ADAT – MIDI
TC Electronic	M350	Twenty-four bits – 256 presets – 99 user slots – 15 reverbs + 15 types of effect – VST plugin – MIDI – S/PDIF

Table A4.2. *Examples of modern multieffects racks*

Glossary

A

Autotune: Tone correction software developed by Antares Audio Technologies.

Auxiliary: On a mixing desk, each input channel has one or more outputs labeled "auxiliary". These can usually be toggled between "postfader" and "prefader" modes (see the corresponding entries later in the glossary) to choose whether the signal should be taken from before or after the fader. The fader controls the output volume of each input channel (track).

B

Bus: Temporary processing channel that allows several channels or inputs to be combined on a mixing desk.

Bus master: Combines all output channels of a mixing desk. In the studio, a copy of the bus master is typically sent to the monitor to allow the person managing the mix to listen in.

Bypass: Bypass mode allows an audio signal to pass through processing equipment without altering its sound characteristics.

C

Cardioid: Term that describes the directionality of certain types of microphone. Cardioid microphones are highly sensitive to sound from the front, but much less sensitive to sound from the rear. They are used to isolate sounds from unwanted ambient noise, which makes them extremely valuable for recording in noisy environments.

Coincident microphones: Pair of cardioid microphones of same type coupled together for sound recording. The sensors (capsules) of the microphones are placed as close as possible (less than 40 cm) at a relative angle between 90° and 135°, pointing toward the sound source.

Compressor: Hardware or software that applies dynamic range compression.

Crosstalk: Interference between two audio signals.

D

DAW (Digital Audio Workstation): Workstation (on a computer or tablet) for recording, editing, mixing, manipulating, creating, processing and reading audio content. Examples of DAWs include Avid Pro Tools, MOTU Digital Performer, Magix Samplitude, Cakewalk Sonar, Apple Logic Pro, PreSonus Studio One and many others.

Decibel (dB): Unit used in acoustics and electronics, defined as 10 times the base-ten logarithm of the ratio of two powers.

De-esser: Tool for reducing whistles and occlusives (stops) in a sound.

Distortion: Undesirable alteration of an audio signal.

Dithering: Technique for improving digital sound data.

Doppler effect: Frequency shift in a sound wave caused by a variation in the distance from the source of emission over time.

DSD (Direct Stream Digital): Protocol for storing and recording audio signals on a digital medium, for example used by Super Audio CDs (SACDs).

DSP (Digital Signal Processing): Processor dedicated to and optimized for executing signal processing tasks (filtering, extraction, conversion, encoding, decoding, compression, etc.).

Dynamic range compression: Audio technique that reduces the difference in level between the loudest and quietest sounds of a signal.

E

Enhancer: Hardware or software tool for improving the sound rendering.

Equalizer: Hardware or software system for correcting the timbre of a sound. Equalizers can filter, boost and cut specific frequency bands of the audio signal. They are commonly used for sound recording, mixing and sound reinforcement.

Exciter: see enhancer.

Expander: Hardware or software tool for increasing (expanding) the dynamic range of an audio signal.

F

Fuzz: Sound effect that creates saturation in the sound signal by means of heavy clipping.

H

Harmonizer: Hardware or software tool that enhances vocals by duplicating them at higher or lower musical intervals (third, fifth, etc.), creating chords.

I

Insert: Integrated access point on the mixing console or desk that allows users to add new devices (peripherals) into the audio flow between the pre-amp and the mixing bus.

L

Leslie: Common name for an amplification cabinet with one or more rotating speakers designed to create a Doppler effect in the sound signal, usually a signal originating from an electronic organ. Named after their inventor, Donald Leslie, these speakers can also be used with instruments other than organs. Effects that simulate the sound of Leslie speakers are often called Leslie effects, "rotovibe," "rotary" or "univibe".

Leslie speakers: see Leslie.

LFO (Low-Frequency Oscillator): Oscillator specially designed for low-frequency variations, often used for tremolo and vibrato effects.

Limiter: Tool for processing the dynamics of an audio signal. Limiters allow us to optimize the overall volume of a sound message, perfectly complementing the possibilities offered by a compressor.

Loudness: Digital quantity representing the volume as a sound as perceived by a human observer. Loudness is a complex psychoacoustic concept that depends on a large number of parameters.

M

Main bus: Primary output bus from a mixing desk. Usually stereo.

Mastering: Final stage of audio postproduction during which the sound rendering of a mix is optimized and calibrated in order to be transferred (for example to a medium like a CD), used or played.

Matrixing: Another term for mastering (see above).

Max/MSP: Music software for synthesizing sound, controlling instruments, recording, performing audio analysis and so on. It was developed in France by IRCAM (Institute for Research and Coordination in Acoustics/Music) in the 1980s.

MIDI (Musical Instrument Digital Interface): Protocol, communication standard and file format (MIDIfile) created for electronic musical instruments, officially released in 1983. MIDI connections are based on three 5-pin DIN connectors: input (MIDI in), output (MIDI out) and forwarding (MIDI Thru). Two cables are required for two-way communication between each device in the musical sequence.

Mixing: During postproduction in a recording studio, mixing is the operation that combines multiple sound sources together to construct a single cohesive overall sound. Mixing is also performed in real time and on-site for live events such as concerts.

N

Noise gate: Processing operation that filters unwanted sounds before amplification.

O

Overdrive: Sound effect used to modify the sound of an instrument by increasing the gain of the amplifier to create controlled distortion.

P

PCM (Pulse Code Modulation): Processing operation for digitizing an electrical signal or analog audio. After sampling, each sample is quantified and transformed into a digital code.

Phoneme: Smallest distinguishable element when dividing a spoken message into segments. This term is used in the field of linguistics, most notably in phonology.

Pitch shifter: Hardware or software tool for applying a sound effect that modifies the pitch of a sound signal while keeping the same tempo (without modifying the duration of the sound).

Postfader: On a mixing desk, the auxiliary "postfader" output is usually a dial that copies part of the signal after it has passed through the fader (linear potentiometer) of the corresponding channel (track). The position of the fader modifies the level of the signal sent to the "postfader".

Prefader: On a mixing desk, the "prefader" output copies part of the signal before it passes through the fader (linear potentiometer) of the corresponding channel (track). The level of this auxiliary output is not affected by the position of the fader.

Presence: System for adding trebles and sharpness to a sound. This is usually done directly at the power stage of the amp via a feedback system (part of the input signal received by the amp is fed back into it).

Q

Quadriphonic sound: Also known as "tetraphonic sound". Procedure designed to create a spatialized sound signal by using four independent channels: front right, front left, back right, back left.

R

Re-recording: Technique that involves recording sound messages by adding them to and mixing them with other prerecorded sounds.

Rotary: See Leslie.

Rotovibe: See Leslie.

RS422: Digital transmission standard based on a serial communication protocol.

S

SACD (Super Audio Compact Disc): Optical digital medium that allows music to be stored in very high quality, created by Sony and Philips and first marketed in 1999.

Saturation: Saturation occurs in an electronic circuit whenever the output level does not increase any further when the input level is increased. Saturation is therefore an alteration or deformation of the sound signal, which is said to be distorted (saturation generates distortion).

Sibilance: Whistling caused by strongly accentuated "sq", "sh", soft "ch", "z" and "j" sounds in speech. Also called the "sibilants" or "occlusives" of a sound.

Sidechain: Feature that varies the sound parameters (usually but not always the gain) as a function of an input signal, typically an external input.

S/PDIF (Sony/Philips Digital Interface): Digital interface for transferring digital audio data, also known as IEC 958. This standard was designed by Sony and Philips in 1989. It competes directly with the professional standard AES/EBU (Audio Engineering Society/European Broadcasting Union).

T

Timbre: The set of sound parameters that characterize instruments and voices and which enable us to identify them.

Tremolo: Periodic variation in the volume (and/or pitch) of a sound signal, usually created by a low-frequency oscillator (LFO).

U

Univibe: See Leslie.

USB (Universal Serial Bus): Standard for serial transmission connecting multiple devices to a computer or any other compatible host. USB was designed in the mid-1990s to replace various other communication devices, such as parallel ports, serial ports, SCSI ports, etc. It requires a specific type of cable and has evolved and improved over time. The latest version, USB 3.1, released in 2014, can theoretically achieve 10 Gbits/s.

V

Vacuum tubes: Also called "electronic tubes". Active electronic component usually used for signal amplification, later replaced by semiconductors (transistors, diodes, etc.). Vacuum tubes were invented by Lee De Forest in 1907 (triode).

Vibrato: Periodic variation in the frequency of a sound signal, usually created by a low-frequency oscillator (LFO).

W

Wah-wah: Musical effect used by trumpeters, trombonists and guitarists based on the principle of shifting the sound spectrum. It can be generated mechanically, with a mute, or electronically, usually with a pedal. Sometimes written "wha-wha" or "wa-wa".

Wurlitzer: Brand of a famous electric piano with 64 keys, based on a mechanism with hammers that strike metal blades. It has a very distinctive and unique sound. Many musicians have used it in the past and still (occasionally) do to this day, especially for rock and jazz. It was invented by Ben F. Meissner in the early 1950s, and peaked in popularity from 1960 to 1975. This piano was a serious competitor of the Fender Rhodes piano, although much less widely used.

X

XLR: Type of connector with a circular cross-section used for connecting professional audio and lighting equipment. It can have three to seven pins. The three-pin version is much more commonly used.

Bibliography

[BAY 13] BAYLE J., *Le guide ultime et zen de Max for Live*, Leanpub, 2013.

[CAP 13] CAPLAIN R., *Technique de prise de son*, 6th ed., Dunod, Paris, 2013.

[CAS 07] CASE A., *SoundFX: unlocking the creative potential of recording studio effects*, Focal Press, 2007.

[ERN 15] ERNOULD F., *Le grand livre du home studio – Tout pour enregistrer et mixer de la musique chez soi*, Dunod, Paris, 2015.

[FLE 95] FLEURY P., MATHIEU J.P., *Vibrations mécaniques acoustiques,* Eyrolles, Paris, 1995.

[FOR 86] FORTIER D., *Le mini studio, théorie et pratique*, Editions Fréquences, 1986.

[GIB 97] GIBSON D., *The Art of Mixing*, Mix Books, 1997.

[JOU 99] JOUHANNEAU J., *Notions élémentaires d'acoustique*, Editions Tec & Doc, 1999.

[HUB 13] HUBER D.M., RUNSTEIN R.E., *Modern Recording Techniques*, Focal Press, 2013.

[HUG 05] HUGONNET C., WALDER P., *Théorie et pratique de la prise de son*, Eyrolles, Paris, 2005.

[HUN 14] HUNTER D., *Guitar Amps and Effect for Dummies*, For Dummies, Wiley, 2014.

[IZH 11] IZHAKI R., *Mixing Audio: Concepts, Practices and Tools*, Focal Press, 2011.

[KAT 14] KATZ B., *Mastering audio: The Art and the Science*, 3rd ed., Focal Press, Wiley, 2014.

[KIN 99] KINSLER E.L., FREY R.F, COPPENS A.B. et al., *Fundamentals of Acoustics*, Wiley, 1999.

[LAN 11] LAGRANGE M., Modélisation sinusoïdale des sons polyphoniques, PhD thesis, University of Bordeaux, 2011.

[LYO 12] LYON E., *Designing Audio Objects for Max/MSP*, A-R Editions, 2012.

[MAN 16] MANZO V.J., *Max/MSP/Jitter for Music: A Practical Guide to Developing Interactive Music Systems for Education and More*, 2nd ed., OUP USA, 2016.

[MAT 69] MATRAS J.J., *L'acoustique appliquée*, Presses Universitaires de France, Paris, 1969.

[MER 13] MERCIER D., *Le livre des techniques du son – Tome 3: L'exploitation*, 4th ed., Dunod, Paris, 2013.

[MER 15] MERCIER D., BOURCET P., CALMET M. et al., *Le livre des techniques du son – Tome 1: Notions fondamentales*, 5th ed., Dunod, 2015.

[OWI 13] OWINSKI B., *The Mixing Engineer's Handbook*, Delmar Cengage Learning, 2013.

[PAR 96] PARKER D.J., STARETT R.A., *CD Recordable Handbook*, Pemberton Press, 1996.

[PER 15] PERIAUX B., OHL J.L., THÉVENOT P., *Le son multicanal – De la production à la diffusion du son 5.1, 3D et binaural*, Dunod, 2015.

[PER 14] PERRINE J., Sound Design, *Mixing, and Mastering with Ableton 9*, Hal Leonard Corporation, Milwaukee, 2014.

[PIE 84] PIERCE J.R., *Le son musical*, Belin, Paris, 1984.

[REI 14] REISS J.D., MCPHERSON A., *Audio Effects: Theory, Implementation and Application*, CRC Press, 2014.

[ROA 16] ROADS C., *L'audionumérique – Musique et informatique*, 3rd ed., Dunod, Paris, 2016.

[ROB 14] ROBISON K., *Ableton Live 9: Create, Produce*, Perform, Focal Press, 2014.

[ROT 95] ROTHSTEIN J., *MIDI: A Comprehensive Introduction*, vol. 7, A-R Editions, 1995.

[RUM 02] RUMSEY F., MCCORMICK T., *Son et enregistrement*, Eyrolles, Paris, 2002.

[SCH 73] SCHAEFFER P., *La musique concrète*, Presses universitaires de France, Paris, 1973.

[SCH 77] SCHAEFFER P., *Traité des objets musicaux*, Le Seuil, Paris, 1977.

[SER 11] SENIOR M., *Mixing secrets for the small studio*, Focal Press, 2011.

[TUR 15] TURRIER C., *Le son – Théorie et technologie*, Ellipses, Paris, 2015.

[VIE 14] VIERS R., EDERY T., *Le guide ultime du sound designer: Comment créer et enregistrer des effets sonores pour le cinéma et la télévision*, Editions Dixit, Paris, 2014.

[WHI 93] WHITE P., L'isolation, *la correction acoustique et le monitoring*, ACME, 1993.

[WHI 93] WHITE P., *Effets et processeurs*, vols 1–2, ACME, 1993.

[WYN 00] WYNER J., *Audio Mastering: Essential Practices*, Berklee Press Publications, Boston, 2000.

Web links

By their very nature, Internet links are unreliable. Resources can be moved to new locations, or sometimes removed completely. All of these links were valid when this book was written. If you find that a link no longer works, you may be able to find an alternative location by using Google or your preferred search engine.

Miscellaneous

Audio restoration: http://www.audio-maniac.com/?page_id=190

Audio restoration: http://www.clickrepair.net/

Amplifier classes: http://www.electronics-tutorials.ws/amplifier/amplifier-classes.html

Audio amplifier classes: http://www.audioholics.com/audio-amplifier/amplifier-classes

Digital amplification, analog: http://www.sonopourtous.fr/fr/article/12-amplification-numerique-analogique-qu-est-ce-que-c-est-

How a digital amplifier works: http://www.axiomaudio.com/blog/digital_amplifier

What types of modification can you find in a distorted signal?: http://www.tpe-distorsion2.sitew.com/2_Les_types_de_distorsions.D.htm#2_Les_types_de_distorsions.D

Audio distortion measurements: http://www.bksv.com/media/doc/BO0385.pdf

The phase vocoder, a tutorial: https://www.eumus.edu.uy/eme/ensenanza/electivas/dsp/presentaciones/PhaseVocoderTutorial.pdf

Introduction to computer music: Phase vocoding: http://iub.edu/~emusic/etext/synthesis/chapter4_pv.shtml

Download links for the "sine sweep": http://www.audiocheck.net/testtones_sinesweep20-20k.php

Software publishers

3R Audio : www.3raudio.com

Ableton: www.ableton.com

Acon Digital Media: acondigital.com

Audacity: audacity.fr/

Audio Damage: www.audiodamage.com

Audio Ease: www.audioease.com

Avid: www.avid.com

Brainworx: www.brainworx-music.de

Cycling'74: www.cycling74.com

D16 Group: d16.pl

Digidesign: www.avid.com

Eventide: www.eventideaudio.com

FabFilter: www.fabfilter.com

Flux: www.fluxhome.com

Focusrite: focusrite.com

IK Multimedia: www.ikmultimedia.com

Izotope: www.izotope.com

Kjaerhus Audio: www.vst4free.com/index.php?dev=Kjaerhus_Audio or www.kvraudio.com/developer/kjaerhus-audio

Lexicon: www.lexicon.com

Magix: www.magix.com

Max for Cats: maxforcats.com

MeldaProduction: www.meldaproduction.com

MOTU: www.motu.com

Nomad Factory: www.nomadfactory.com

PSP Audioware: www.pspaudioware.com

Rob Papen: www.robpapen.com

Softube: www.softube.com

Sonic Timeworks: www.sonictimeworks.com

Sonar: www.cakewalk.com

Sonnox: www.sonnox.com

SoundToys: www.soundtoys.com

Spectrasonics: www.spectrasonics.net

Steinberg: www.steinberg.net

TC Electronic: www.tcelectronic.com

Universal Audio: www.uaudio.com

Valhalla: valhalladsp.com

Voxengo: www.voxengo.com

Waves: www.waves.com

Wave Arts: wavearts.com

Sound effects

Vintage Digital website dedicated to vintage sound effects: http://www.vintagedigital.com.au/

Distortions: The challenges of sound reproduction: http://www.lesnumeriques.com/labo-son-test-audio-se-passe-comment-a1166/les-distorsions-eternel-probleme-de-la-reproduction-sonore-ap793.html

Noise distortion and total harmonic distortion: http://www.audiosonica.com/fr/cours/post/204/Bruit-Distorsion_et_distorsion_harmonique_totale_THD

The invention of wah-wah pedal: http://priceonomics.com/the-invention-of-the-wah-wah-pedal/

Acoustics of rooms: http://physique.unice.fr/sem6/2012-2013/PagesWeb/PT/Reverberation/page1.html

History of reverb: http://www.rts.ch/couleur3/8498599-l-histoire-de-la-reverberation-du-son-.html

What is reverb? http://www.mediacollege.com/audio/reverb/intro.html

The ultimate guide to effects, reverb: http://www.musicradar.com/tuition/tech/the-ultimate-guide-to-effects-reverb-461487

Reverb in audio and music production: http://www.sonible.com/blog/reverb-audio-production/

Convolution reverb: http://www.sounddesigners.org/articles/theorie/item/14-la-r%C3%A9verb%C3%A9ration-par-convolution.html

Convolution reverb explained: http://www.bhphotovideo.com/find/newsLetter/Convolution-Reverb.jsp/

How to use convolution reverb and effects: http://music.tutsplus.com/tutorials/how-to-use-convolution-for-reverb-effects--audio-1089

Reverberation time: www.bksv.com/media/doc/BO0228.pdf

Echo and reverberation: http://personal.ee.surrey.ac.uk/Personal/P.Jackson/ee1.lab/D3_echo/D3_EchoAndReverberationExpt.pdf

Make your guitar sound "shimmer": http://www.soundonsound.com/techniques/make-your-guitar-sound-shimmer

Building a shimmer reverb with Reason: http://www.propellerheads.se/blog/building-shimmer-reverb

Pedal tricks: The shimmer effect: http://pedals.thedelimagazine.com/pedal-tricks-the-shimmer-effect/

How to use convolution for reverb and effects: http://music.tutsplus.com/tutorials/how-to-use-convolution-for-reverb-effects--audio-1089

Convolution reverb explained: http://www.bhphotovideo.com/find/newsLetter/Convolution-Reverb.jsp/

Autotune by Antares: http://www.antarestech.com/

Hardware manufacturers

Alesis: www.alesis.com

Behringer: www.music-group.com/brand/behringer/home

Boss: www.boss.info

Cedar: www.cedar-audio.com

Dunlop: www.jimdunlop.com

Electro Harmonix: www.ehx.com

Fender: www.fender.com

Gibson: www.gibson.com

Hammond: www.hammondorganco.com

Ibanez: www.ibanez.com

Leslie: www.hammondorganco.com/products/leslie/

Lexicon: www.lexiconpro.com

Line 6: www.line6.com

Mooer: www.mooeraudio.com

Moog: www.moogmusic.com

MXR: www.jimdunlop.com

Musitronics Mu-Tron: www.mu-tron.org

Pultec: www.pulsetechniques.com

Rocktron: www.rocktron.com

Vox: www.voxamps.com

Weiss: www.weiss.ch

Yamaha: www.yamaha.com

Audio filtering

Audio filters: http://www.sonelec-musique.com/electronique_bases_filtre_audio.html

Audio filters: http://www2.ece.ohio-state.edu/~anderson/Outreachfiles/AudioEqualizerPresentation.pdf

Sound synthesis and modeling software

Website of the publisher of Max/MSP: http://cycling74.com/products/max/

CodeLab forums Section dedicated to Max/MSP: http://codelab.fr/max-msp

CodeLab forums – Section dedicated to PureData: http://codelab.fr/pure-data

CodeLab forums – Section dedicated to SuperCollider: http://codelab.fr/supercollider

List of Max objects: http://maxobjects.com/

Max/MSP patches by Florian Gourio: http://www.floriangourio.fr/patches-max-msp/

The Pure Data community (downloads and more): http://puredata.info/

Programming with Pure Data: http://www.flossmanualsfr.net/puredata/

Website for downloading jMax: http://sourceforge.net/projects/jmax/

SuperCollider website: http://supercollider.github.io/

Native Instruments and its Reaktor software website: http://www.native-instruments.com/fr/products/komplete/synths/reaktor-6/

General-purpose websites

Activstudio: http://www.activstudio.fr/mixage-audio/

Attack magazine, a magazine dedicated to music and digital audio: https://www.attackmagazine.com/

Audiofanzine, website dedicated to musical audio hardware: http://www.audiofanzine.com

Canford, seller audio and video products: http://www.canford.fr/

Training and resources for sound enthusiasts: http://deveniringeson.com/

Harmony Central, one of the best websites dedicated to musical audio: www.harmonycentral.com

Keyboard magazine: http://www.keyboardmag.com/

KR Home studio, the magazine for musical creation: https://www.kr-homestudio.fr/

KVR, a website for free audio plugins, among others: https://www.kvraudio.com/

Setting up a Home Studio: http://www.monter-son-home-studio.fr/

Mastering addict: http://mastering-addict.com/#sthash.yun06RU3.dpbs

Music Store Professional, hardware seller: http://www.musicstore.de/en_OE/EUR

MusicTech, website dedicated to sound engineers and music production: http://www.musictech.net/

Musiker Board, a website for musical audio (in German): www.musiker-board.de

Pro Audio Review magazine: http://www.prosoundnetwork.com/article.aspx?articleid=39995

ProSound: http://www.prosoundnetwork.com/

SoundClick, website dedicated to musical audio: www.soundclick.com

Synthmuseum, website dedicated to old synthesizers: http://www.synthmuseum.com/

Synthtopia, a museum of old synthesizers: http://www.synthtopia.com/

Tape Op magazine : http://tapeop.com/

Thomann, online sales platform for musical products: http://www.thomann.de/fr/index.html

Website dedicated to vintage synthesizers: http://www.vintagesynth.com/

VST4free, website for free VST plugins: http://www.vst4free.com/index.php?plugins=Synthesizers

Woodbrass, online sales platform for musical products: http://www.woodbrass.com/

Zikinf, general purpose website for music and audio: http://www.zikinf.com/

Multichannel sound

Dolby digital 5.1: http://www.dolby.com/us/en/technologies/dolby-digital.html

Dolby Surround 7.1: http://www.dolby.com/us/en/technologies/dolby-surround-7-1.html

DTS : http://dts.com/

News about multichannel audio: http://www.lesonmulticanal.com/

Recording multichannel sound: http://multiphonie.free.fr/docs/multicanal_jm_lhotel.pdf

Thoughts about 5.1 multichannel sound recording and postproduction: http://www.conservatoiredeparis.fr/fileadmin/user_upload/Recherche/Le_service_audiovisuel/post-prod-multicanal5-1.pdf

Multichannel recording: http://wiki.audacityteam.org/wiki/Multichannel_Recording

Multichannel sound recording: http://microphone-data.com/media/filestore/articles/MMAD-10.pdf

Surround sound by Dolby: http://www.dolby.com/us/en/technologies/surround-sound.html

DTS, DTS-HD Master audio: http://www.hdfever.fr/2009/03/09/digital-theater-system-dts-hd-master-audio/

The DTS website: http://dts.com/

Theory of sound

Physics theory of sound and perception: http://www.sonorisation-spectacle.org/physique/son.html

Long term sinusoidal modeling of speech: http://tel.archives-ouvertes.fr/tel-00211294/document

Understanding Pitch-Synchronous Overlap-Add (PSOLA): http://dafx.org/understanding-pitch-synchronous-overlap-add-psola/

Time scaling (OLA – SOLA – PSOLA): ftp://ftp.esat.kuleuven.be/sista/jszurley/peno/timescaling.pdf

Fourier transforms and theirapplications: http://www.techniques-ingenieur.fr/base-documentaire/sciences-fondamentales-th8/applications-des-mathematiques-42102210/la-transformee-de-fourier-et-ses-applications-partie-1-af1440/

Constructing and analyzing a signal by Fourier analysis: http://www.tangentex.com/AnalyseFourier.htm

Sine sweep vibration testing for modal response primer: http://www.noar.technion.ac.il/images/attachments/Scitech/Eric_Sauther_Sine_Sweep_Tutorial.pdf

Video tutorials about audio

Elephorm: http://www.elephorm.com/audio-mao.html

Mj tutoriels: http://www.mjtutoriels.com/18-techniques-audio

Tutorom: http://www.tutorom.fr/categories-de-tutoriels/fr/audio

Virtual Production School (VPS): http://www.tutoriels-mao.com/les-tutoriels/mix-et-master-de-a-%C3%A0-z-avec-des-plugins-gratuits-detail

Learning to mix with Ableton: http://le-son-ableton.fr/apprendre-le-mixage-avec-ableton/

Anto's tutorials: http://www.tutodanto.com/c/voir-tous-les-tutos-mao

Tuto.com: http://fr.tuto.com/tuto/audio-mao/

Home studios for beginners: http://www.home-studio-debutant.com/

Index

33 rpm, 46
45 rpm, 46
78 rpm, 46
78 rpm, 46

A

absorption, 39
  coefficient, 39
AC-3, 51
accentuation, 96, 97
acoustic
  duct, 206, 216, 217
  impedance, 39
  pressure, 14
  sound pressure level, 4
acoustic reflex, 272
acoustics, 3
activation, 76
Adaptative Transform Acoustic
  Coding (ATRAC), 59
Adobe Audition CC, 231
algorithm, 208, 212
Allgemeine Elektricitäts-Gesellschaft
  (AEG), 46
Altiverb, 207
ambience, 62, 197, 225
Ampex, 106
amplitude, 7, 23, 24, 129, 159. 262
  modulation, 127

analog
  digital conversion, 267
  multitrack tape recorder, 253
angle of
  incidence, 37, 40
  reflection, 37, 40
antiphase, 125
aperiodic, 21
apex, 12
argument, 82
Artificial Double Tracking (ADT),
  106
attack, 100, 156–158, 169, 171–174,
  179, 180, 184, 185
attenuation, 81, 216
  range, 185
audibility, 3
audio
  Antares Technologies, 144
  CD, 43
  filtering, 182
  flow, 130
  processing chain, 95
audiocheck, 230
auditory
  nerve, 13
  spatial awareness, 15
aural exciter, 192
auricle, 9

auto-tune, 135, 142, 145, 146
auto-wah, 100
Avid Protools, 63

B

background noise, 180, 267
Bäder, Karl Otto, 205
balance, 129, 171
band-stop, 81
bandwidth, 47, 83, 217, 251
basilar membrane, 12
bass, 84
Baxandall, 83
bel, 5
bell, 44, 83
Berliner, 44
Berry, Chuck, 188
binaural, 14
Blackmore, Ritchie, 188
Blu-ray, 43, 54
Bode plot, 82
bone conduction, 14
brickwall, 175
bundle, 110, 113, 116
Burnette, Johnny, 188
bypass, 96

C

cardioid, 125
CD, 276
ceiling, 176
celerity, 4, 121
center width, 53
chamber, 225
chorus, 105, 115, 116, 130, 249
clavinet, 104
clicks, 266
clipping, 175, 187, 267
close threshold, 181
cochlea, 11, 12
cocktail party, 20
coincident microphones, 47
combo, 119

complex sound, 31
component, 28
compression, 122, 156, 184, 277
   dynamic range, 157
   invisible, 170
   serial, 171
   with a sidechain 171
compressors, 156, 158, 179, 186
   field effect transistor (FET), 165
   multiband, 166
   operational transconductance amplifier, 165
   optical, 165
   tube, 165
   voltage controlled amplifier (VCA), 165
cry-baby, 98
cut-off, 223
concave, 38
constructive interference, 33, 34
continuous, 22
convexe, 38
convolution, 208, 215
Cros, Charles, 44
cross-talk, 62, 189
crossover, 102, 221
crunch, 187, 188
cut-off release, 223

D

damping, 28, 210, 221, 241
dB(A), 5, 6
de-esser, 96, 184
DeArmond, 262
debuzzer, 266, 268
decay, 221, 223
   level, 221
   time, 221
Decca tree, 63
decibel, 5, 6
declicker, 266
declipper, 266, 267

deconvolution, 215, 227, 230, 232–234, 237
decrackler, 266, 267
Deep Purple, 188
dehummer, 268
delay, 103, 108, 139, 184, 197, 216, 218, 243, 252, 256, 272, 277
  analog, 244
  lines, 212
  stereo, 251
  tape, 244
denoiser, 266–268
densification, 213
density, 222
depth, 109, 129, 132
Descartes' law, 37
destratcher, 266
destructive, 33
  interference, 34
diffraction, 35, 36
diffuse reverb, 15, 226
diffusion, 38, 213, 221, 222, 241
diffusor, 200, 201
digital, 245
  signal processing, 212
  signal processor, 107
  theater system, 48, 55
Digital Audio Tape (DAT), 276
Digital Audio Workstation (DAW), 63, 77, 147, 152, 155, 178
dilation, 122
dimension, 53
direct sound, 20
Direct Stream Digital (DSD), 67
  DSD-CD, 69
dispersion, 32
distortion, 187
dither, 177
dithering, 177
Dolby, 63
  A, 48
  Atmos, 55
  Digital 5.1, 48
  Digital AC-3, 51

Digital Plus, 54
Headphone, 58
SR, 50
SR-D, 52
Stereo, 49
Surround, 48, 50, 53
Surround EX, 52
Surround Pro-Logic, 48, 50
Surround Pro Logic II, 52
TrueHD, 54
Virtual Speaker, 59
Doppler, 120
double ORTF, 63
doubling, 251
dry, 107, 220, 223
dry/wet, 238, 251
DSD, 67
DSP, 107, 212
DTS, 48, 55, 63
  96/24, 57
  ES 6.1, 57
  HD High Resolution Audio, 58
  HD Master Audio, 57
  Headphone, 59
  Interactive, 59
  Neo 6, 56
  X, 58
Dudley, Homer, 143
Dussaud, Francois, 44
DVD, 43, 51
  A, 65, 70
  Audio, 48, 62, 65
  HD DVD, 43, 54
dynamic equalization, 192
dynamics, 79, 126, 155, 170, 187

E

ear, 9
  canal, 18
  inner, 10
  middle, 10
  outer, 20
early reflections, 221

echo, 28, 198, 218, 243, 256
　chamber, 216, 243
Edison, Thomas, 44
effect(s), 245
　Doppler, 40, 120
　dynamic, 71, 155
　filtering, 71, 81
　frequency, 71, 131
　fuzz, 187
　Haas, 20
　modulation, 105
　precendence, 20
　rotary, 40
　time, 71, 197
elastic limit, 4
elasticity, 4
electric
　field, 2
　piano, 263
electronic tubes, 211
ElektroMessTecknik (EMT), 203
ELP, 188
Emulator, 77
Endolymph, 12
enhancer, 192, 271
envelope, 28, 174
EQ, 84
equalization, 82, 84, 96, 104, 216, 251, 277
equalizer (EQ), 83, 84, 91, 103, 249, 271, 186
　band-stop, 86, 88
　dynamic, 86, 90
　graphic, 86
　linear phase, 86, 89
　parametric, 86, 87, 94
　semi-parametric, 86, 88
eustachian tube, 10
exciter, 192, 193
expander, 178
external
　input, 175, 184
　signal, 182

F

Falcon, 264
feedback, 107, 109, 245, 251
Fender, 76, 264
filter, 110, 113, 166, 246
　band-pass, 81, 88, 111
　high-pass, 81, 172
　low-cut comp, 162
　low-pass, 81, 162
　resonant, 100
　　band-stop, 81
　　digital, 100
Fine, Bill, 200
flanger, 105, 106, 130
flanging, 106, 109
Fletcher-Munson, 14
focusing, 38
Ford, Mary, 243
Formant, 149
Fourier transform, 23
free-field, 215
Freesound, 230
frequency, 7, 23, 213
　attenuation, 223
　bands, 46, 84, 111
　beat, 41
　central, 99
　cut-off, 88
　decay, 223
　detector, 185
　distribution, 13, 25
　fundamental, 23, 24, 27, 29
　range, 185
　response, 189
　sampling, 236
Fripp, Robert, 244
fuzz, 187, 257, 258, 272
fuzzwha, 257

G

gain, 82, 88, 95, 110, 157, 170
gate, 178, 223

decay, 223
gated, 225
germanium transistor, 189
Gibson, 264
graphic, 24

H

hair cells, 13
hall, 225
Hamasaki square, 63
Hammond, 117, 188
Hammond, Laurens, 118, 203
Hammond-Suzuki, 120
harmonic
   distortion, 189, 195
   exciter, 192
harmonics, 108
harmonizer, 135
Head Related Transfer Functions (HRTF), 18
headset, 170
Hendrix, Jimi, 106, 136, 188
Hertz (Hz), 7
hi-fi, 45, 201
high-fidelity, 45, 47, 201
high-midrange, 84
Hildebrand, 142
hissing, 184
hold, 179, 181, 246
Holman, 60
holophone H2, 63
home-cinema, 102
Huygens, Christian, 36
Huygens' principle, 36
hysteresis, 181

I, J

I'm T-Pain, 154
impulse response (IR), 215, 227, 230, 232, 234, 237
   Utility, 234, 239
Incus, 10
initial state, 90

insert, 185
intensity, 3, 41
interaural level, 15
Interaural Level Difference (ILD), 18
Interaural Time Difference (ITD), 15, 18
interference, 38, 40
intial level, 221
isosonic curves, 14
ITD, 15, 18
Johnson, Eldridge, 44

K

K7 audio, 43
Kendrick, Warren, 106
Key, 175, 184, 186
kinetic energy, 211
knee, 160, 169, 171–174, 179
   soft, 160
   hard, 160
Kraftwerk, 143
Kruesi, John, 44
Kuhl, Walter, 203

L

laserdisc, 43, 51
law of
   reflection, 37
   the first wavefront, 20
Les Paul, 106, 243
Leslie, 117
   speaker, 117, 122
Lexicon, 206
limiter, 103, 161, 175, 179
link, 161
Lioret, Henri, 44
loop, 246, 249
   effect, 273
loopers, 271
loudness, 4, 27, 155
   contour, 14
low-midrange, 84
low-pass, 172

low frequency
  effect, 51, 56, 60, 162
  oscillator, 100, 107, 128, 132, 246, 249
Lucasfilm, 52, 60

M

magnetic
  field, 2
  plate, 216
  tape, 216
  tape recorders, 46, 243
major mode, 151
make-up gain, 157, 160, 170, 173, 174
malleus, 10
masking, 171
mass, 30
mastering, 63, 89, 104, 175, 177
maxima, 63
Mercury Records, 200
Meridian Lossless Packing (MLP), 65
metronome, 272
Meyer-Eppler, Werner, 143
MicDroid, 153
microcomputers, 271
MIDI, 74, 246
mids, 84
mini-disc, 59
mix, 108, 162, 223, 245
mixing, 72, 104, 166, 169, 173, 177, 180
modulation speed, 108
modulator, 127
module, 82
monitoring, 170
monophony, 138
Moore, Gary, 188
morphology, 18
Morse, Steve, 188
Movie, 53, 56
MP-Matrix, 48

Multiphonic Microphone Array Design (MMAD), 63
Music, 53, 56
mute, 98
Mu-Tron III, 100

N

necklace, 204
Neumann, Georg, 45
New York compression, 170
Newton, 4
nodal
  group, 31
  sound, 31
noise, 23
  gate, 180
nonlinear, 225
notch, 81, 86, 89

O

Octafuzz, 257, 258
Octave, 84, 135, 137
Octaver, 135
Octavia, 136, 258
open threshold, 181
operational amplifiers, 100
Optimized Cardioid Triangle (OCT), 63
order, 81
organ of Corti, 12
organists, 188
output level, 110, 176, 177
oval window, 10, 12
overdrive, 75, 187, 188, 274
overdubs, 272

P

Pa, 4
pan-delay, 252
panning, 248
panorama, 53
parallel, 170

pascals, 4
pedal, 71, 74, 247, 261
pedalboards, 75
perilymph, 12
periodic, 21
periodicity, 21
Pfleumer, Fritz, 4
pharynx, 10
phase, 32, 34, 36, 82
   inversion, 39
   rotation, 89
   shift, 89
   variation, 242
   vocoder, 276
phaser, 105, 111, 130, 249
Philipps, Sam, 243
phoneme, 152
phonograph, 47
piano, 85
ping-pong delay, 252
Pink Floyd, 188, 244
pipe, 216
pitch, 3, 7, 115, 131
   correction, 150
   scaling, 276
pitch-shifter, 135, 137, 140
plate, 208, 224
plugin, 63, 71, 77, 107, 155, 161,
   207, 213, 221, 245, 248, 239
point of
   audition, 32
   rotation, 162
Polsfuss, Lester William, 106
polyphonic, 138
pop, 26
   filter, 126
post-production, 63, 148
Poulsen, Valdemar, 45
pre-delay, 220, 240
presence, 96
preset, 224
principle of superposition, 35
pseudo-periodic, 26
psophometric, 6

psychoacoustic, 3, 27, 30
Pulse
   Code Modulation (PCM), 65
   Width Modulation (PWM), 100
pumping, 169
pure sound, 31
Putnam, Bill, 202

Q, R

quadriphonic, 47
Q-wah, 100
rack, 72, 247
range, 28, 84, 85, 116
   fundamental, 28
   spectral, 28
   vocal, 186
rate, 132
ratio, 157, 161, 169, 171–175
re-recording, 276
read, 169
read head, 254
recording, 62, 72
   head, 254
reflex latency, 10
reflections, 37, 39, 211, 221, 227
   primary, 220, 221
   secondary, 221
refraction, 39
regen, 109
rejection filter, 81, 89
relâchement, 156, 157, 159, 169,
   171–174, 176, 179, 180, 184, 185
release, 157, 159
resampling, 276
resonance, 242
reverb, 28, 39, 75, 184, 197, 255,
   274, 277
   algorithmic, 212
   analog, 208
   artificial, 202, 204
   chamber, 200, 202
   convolution, 207
   digital, 205, 208, 215

magnetic tape, 206
plate, 203
processors, 207
software, 207
spring, 119
synthetic, 212, 213
tail, 226
time, 198, 220, 226
reverse, 216, 225, 272
ring modulation, 105, 127, 130
rise time, 132
room size, 226, 241
rotary, 105, 117, 123, 130
rotating
  drum, 120
  horn, 120
rotation point compressor, 162
rotovibe, 105, 117
RS422, 74
RT60, 220

S

S/PDIF, 57, 74
Sabine, 39, 199
Sabine's laws, 200
sample, 246
samplers, 271
sampling depth, 177
Santana, Carlos, 188
saturation, 187, 189, 242, 267
Schaeffer, Pierre, 30
semitone, 137, 140
send, 273
sensors, 12, 209
  and transducers, 211
Shannon, Claude Elwood, 9
shape, 132
shelf, 92, 96
shimmer, 257, 259
sibilance, 96, 184, 186, 187
sidechain, 161, 162, 171, 175, 179, 182, 184, 186

external, 172
internal, 172
signal
  -to-noise ratio, 189
  impulse, 222
  processing, 107
sine
  sweep, 215, 227, 228, 230, 235, 237
  wave, 108, 109, 127, 128, 132, 215, 292, 293, 294
slapback, 243, 251
slave, 161
slope, 28, 81
  rate, 81
smoothening the sound, 173
software component, 71
SOLA, 276
Songify, 153
Sonogram, 26
Sonovox, 143
Sony Dynamic Digital Sound (SDDS), 59
Sound, 99, 113
  cohesion, 172
  intensity, 38
  level, 4
  propagation, 37
  reinforcement, 72
  restoration, 266
  signature, 215
  spectrum, 24
  thickness, 173
Soundfield, 63
Space designer, 234, 239
spatial localization, 62
spatialization, 277
spatializers, 197, 216, 277
spectral
  analysis, 1, 24
  envelope, 25
  recording, 50
  sine modeling, 276

Index   345

spectrogram, 26
spectrum, 18, 102, 186
　analyzer, 83
　comb, 25
　continuous, 26
　discontinuous, 25
　line, 25
speed, 132
　of sound, 4
spread, 226
spring, 208, 211, 225
stapes, 10
　muscle, 10
static noise, 267
stereo
　image, 226
　offset, 109
stereocilia, 13
Stockhausen, Karl, 127
Storytone, 263
sub-bass, 84
subwoofer, 51
Super
　Audio CD (SACD), 48, 62, 67, 70
　High Material (SHM) SACD, 69
surround, 49, 60, 103
sustain, 187, 222
syllable, 152
Synchronous OverLap and Add, 276
synthesizer, 130, 259

T

Tainter, Charles, 44
tap-tempo, 246, 249, 252
tape recorder, 271
target state, 90
tectorial membrane, 13
telegraphone, 45
tempo, 252
The Rolling Stones, 188
threshold, 90, 156, 157, 160, 161,
　　169, 171–174, 176–178, 180,
　　185

THX, 52, 60
　certification, 59
　I/S Plus, 60
　select, 61
　ultra 2, 61
timbre, 1, 3, 8, 18, 27, 84, 158, 160
time
　code, 56
　interval, 222
　stretching, 275
tonal group, 31
tone, 132
　cabinet, 118
tonotopy, 13
transducer, 208, 211, 212
transient phenomena, 27
transients, 215, 227, 239
　attack, 17
　release, 27, 28
transmittance, 82
transposers, 135
treble, 84
tremolo, 276
Tremolux, 264
triangle wave, 109, 128, 132
tuning, 115
T-wah, 100
tympanic duct, 12

U, V

univibe, 105, 117, 123
USB, 74, 246
vacuum tubes, 156, 187
Van Halen, Eddie, 188
variation signal, 132
vestibular duct, 12
vestibule, 11
VHS, 53
vibrato, 27, 75, 131, 150, 151, 262,
　　264
Vibrolux, 264
video disc, 43
vinyl, 46

vocal
  processors, 142, 145
  synthesis, 142
vocoder, 143, 147
voder, 143
voice contours, 148
voloco, 154
voltage-controlled amplifier (VCA), 165
volume, 262
Vox
  Continental, 98
  Wah-wah, 98

W, Z

wah-wah, 97, 99, 104, 257
wave, 209
  diffracted, 35
  form, 109
  incident, 38
  mechanical, 2
  plane, 32
  refracted, 38, 39
  secondary, 35
  spherical, 32
wavefronts, 32, 38
wavelength, 7, 8, 36, 121
Waves IR-1, 231
wet/dry, 245
white noise, 22, 31, 267
width, 109, 129, 226
Wray, Link, 188
Zappa, Frank, 244

Other titles from

in

Waves

2017

DAHOO Pierre-Richard, LAKHLIFI Azzedine
*Infrared Spectroscopy of Diatomics for Space Observation
(Infrared Spectroscopy Set – Volume 1)*

PARET Dominique, HUON Jean-Paul
Secure Connected Objects

PARET Dominque, SIBONY Serge
Musical Techniques: Frequencies and Harmony

STAEBLER Patrick
Human Exposure to Electromagnetic Fields

2016

ANSELMET Fabien, MATTEI Pierre-Olivier
Acoustics, Aeroacoustics and Vibrations

BAUDRAND Henri, TITAOUINE Mohammed, RAVEU Nathalie
The Wave Concept in Electromagnetism and Circuits: Theory and Applications

PARET Dominique
Antennas Designs for NFC Devices

PARET Dominique
Design Constraints for NFC Devices

WIART Joe
Radio-Frequency Human Exposure Assessment

2015

PICART Pascal
New Techniques in Digital Holography

2014

APPRIOU Alain
Uncertainty Theories and Multisensor Data Fusion

JARRY Pierre, BENEAT Jacques N.
RF and Microwave Electromagnetism

LAHEURTE Jean-Marc
UHF RFID Technologies for Identification and Traceability

SAVAUX Vincent, LOUËT Yves
MMSE-based Algorithm for Joint Signal Detection, Channel and Noise Variance Estimation for OFDM Systems

THOMAS Jean-Hugh, YAAKOUBI Nourdin
New Sensors and Processing Chain

TING Michael
Molecular Imaging in Nano MRI

VALIÈRE Jean-Christophe
Acoustic Particle Velocity Measurements using Laser: Principles, Signal Processing and Applications

VANBÉSIEN Olivier, CENTENO Emmanuel
Dispersion Engineering for Integrated Nanophotonics

2013

BENMAMMAR Badr, AMRAOUI Asma
Radio Resource Allocation and Dynamic Spectrum Access

BOURLIER Christophe, PINEL Nicolas, KUBICKÉ Gildas
Method of Moments for 2D Scattering Problems: Basic Concepts and Applications

GOURE Jean-Pierre
Optics in Instruments: Applications in Biology and Medicine

LAZAROV Andon, KOSTADINOV Todor Pavlov
Bistatic SAR/GISAR/FISAR Theory Algorithms and Program Implementation

LHEURETTE Eric
Metamaterials and Wave Control

PINEL Nicolas, BOURLIER Christophe
Electromagnetic Wave Scattering from Random Rough Surfaces: Asymptotic Models

SHINOHARA Naoki
Wireless Power Transfer via Radiowaves

TERRE Michel, PISCHELLA Mylène, VIVIER Emmanuelle
Wireless Telecommunication Systems

2012

LALAUZE René
Chemical Sensors and Biosensors

LE MENN Marc
Instrumentation and Metrology in Oceanography

LI Jun-chang, PICART Pascal
Digital Holography

2011

BECHERRAWY Tamer
Mechanical and Electromagnetic Vibrations and Waves

GOURE Jean-Pierre
Optics in Instruments

GROUS Ammar
Applied Metrology for Manufacturing Engineering

LE CHEVALIER François, LESSELIER Dominique, STARAJ Robert
Non-standard Antennas

2010

BEGAUD Xavier
Ultra Wide Band Antennas

MARAGE Jean-Paul, MORI Yvon
Sonar and Underwater Acoustics

2009

BOUDRIOUA Azzedine
Photonic Waveguides

BRUNEAU Michel, POTEL Catherine
Materials and Acoustics Handbook

DE FORNEL Frederique, FAVENNEC Pierre-Noël
Measurements using Optic and RF Waves

FRENCH COLLEGE OF METROLOGY
Transverse Disciplines in Metrology

2008

FILIPPI Paul J.T.
Vibrations and Acoustic Radiation of Thin Structures

LALAUZE René
Physical Chemistry of Solid-Gas Interfaces

2007

KUNDU Tribikram
Advanced Ultrasonic Methods for Material and Structure Inspection

PLACKO Dominique
Fundamentals of Instrumentation and Measurement

RIPKA Pavel, TIPEK Alois
Modern Sensors Handbook

2006

BALAGEAS Daniel *et al.*
Structural Health Monitoring

BOUCHET Olivier *et al.*
Free-Space Optics

BRUNEAU Michel, SCELO Thomas
Fundamentals of Acoustics

FRENCH COLLEGE OF METROLOGY
Metrology in Industry

GUILLAUME Philippe
Music and Acoustics

GUYADER Jean-Louis
Vibration in Continuous Media